Journal of Applied Logics - IfCoLog Journal of Logics and their Applications

Volume 5, Number 7

October 2018

Disclaimer

Statements of fact and opinion in the articles in Journal of Applied Logics - IfCoLog Journal of Logics and their Applications (JAL-FLAP) are those of the respective authors and contributors and not of the JAL-FLAP. Neither College Publications nor the JAL-FLAP make any representation, express or implied, in respect of the accuracy of the material in this journal and cannot accept any legal responsibility or liability for any errors or omissions that may be made. The reader should make his/her own evaluation as to the appropriateness or otherwise of any experimental technique described.

© Individual authors and College Publications 2018
All rights reserved.

ISBN 978-1-84890-280-9
ISSN (E) 2055-3714
ISSN (P) 2055-3706

College Publications
Scientific Director: Dov Gabbay
Managing Director: Jane Spurr

http://www.collegepublications.co.uk

Printed by Lightning Source, Milton Keynes, UK

All rights reserved. No part of this publication may be reproduced, stored in a retrieval system or transmitted in any form, or by any means, electronic, mechanical, photocopying, recording or otherwise without prior permission, in writing, from the publisher.

Editorial Board

Executive Editors
Dov M. Gabbay, Sarit Kraus and Jörg Siekmann

Editors

Marcello D'Agostino
Natasha Alechina
Sandra Alves
Arnon Avron
Jan Broersen
Martin Caminada
Balder ten Cate
Agata Ciabttoni
Robin Cooper
Luis Farinas del Cerro
Esther David
Didier Dubois
PM Dung
Amy Felty
David Fernandez Duque
Jan van Eijck

Melvin Fitting
Michael Gabbay
Murdoch Gabbay
Thomas F. Gordon
Wesley H. Holliday
Sara Kalvala
Shalom Lappin
Beishui Liao
David Makinson
George Metcalfe
Claudia Nalon
Valeria de Paiva
Jeff Paris
David Pearce
Brigitte Pientka
Elaine Pimentel

Henri Prade
David Pym
Ruy de Queiroz
Ram Ramanujam
Christian Retoré
Ulrike Sattler
Jane Spurr
Kaile Su
Leon van der Torre
Yde Venema
Rineke Verbrugge
Heinrich Wansing
Jef Wijsen
John Woods
Michael Wooldridge

Area Scientific Editors

Philosophical Logic
Johan van Benthem
Lou Goble
Stefano Predelli
Gabriel Sandu

New Applied Logics
Walter Carnielli
David Makinson
Robin Milner
Heinrich Wansing

Logic and category Theory
Samson Abramsky
Joe Goguen
Martin Hyland
Jim Lambek

Proof Theory
Sam Buss
Wolfram Pohlers

Logic and Rewriting
Claude Kirchner
Jose Meseguer

Human Reasoning
Peter Bruza
Niki Pfeifer
John Woods

Modal and Temporal Logic
Carlos Areces
Melvin Fitting
Victor Marek
Mark Reynolds.
Frank Wolter
Michael Zakharyaschev

Automated Inference Systems and Model Checking
Ed Clarke
Ulrich Furbach
Hans Juergen Ohlbach
Volker Sorge
Andrei Voronkov
Toby Walsh

Formal Methods: Specification and Verification
Howard Barringer
David Basin
Dines Bjorner
Kokichi Futatsugi
Yuri Gurevich

Logic and Software Engineering
Manfred Broy
John Fitzgerald
Kung-Kiu Lau
Tom Maibaum
German Puebla

Logic and Constraint Logic Programming
Manuel Hermenegildo
Antonis Kakas
Francesca Rossi
Gert Smolka

Logic and Databases
Jan Chomicki
Enrico Franconi
Georg Gottlob
Leonid Libkin
Franz Wotawa

Logic and Physics (space time. relativity and quantum theory)
Hajnal Andreka
Kurt Engesser
Daniel Lehmann
lstvan Nemeti
Victor Pambuccian

Logic for Knowledge Representation and the Semantic Web
Franz Baader
Anthony Cohn
Pat Hayes
Ian Horrocks
Maurizio Lenzerini
Bernhard Nebel

Tactical Theorem Proving and Proof Planning
Alan Bundy
Amy Felty
Jacques Fleuriot
Dieter Hutter
Manfred Kerber
Christoph Kreitz

Logic and Algebraic Programming
Jan Bergstra
John Tucker

Logic in Mechanical and Electrical Engineering
Rudolf Kruse
Ebrahaim Mamdani

Logic and Law
Jose Carmo
Lars Lindahl
Marek Sergot

Applied Non-classical Logic
Luis Farinas del Cerro
Nicola Olivetti

Mathematical Logic
Wilfrid Hodges
Janos Makowsky

Cognitive Robotics: Actions and Causation
Gerhard Lakemeyer
Michael Thielscher

Type Theory for Theorem Proving Systems
Peter Andrews
Chris Benzmüller
Chad Brown
Dale Miller
Carsten Schlirmann

Logic Applied in Mathematics (including e-Learning Tools for Mathematics and Logic)
Bruno Buchberger
Fairouz Kamareddine
Michael Kohlhase

Logic and Computational Models of Scientific Reasoning
Lorenzo Magnani
Luis Moniz Pereira
Paul Thagard

Logic and Multi-Agent Systems
Michael Fisher
Nick Jennings
Mike Wooldridge

Logic and Neural Networks
Artur d'Avila Garcez
Steffen Holldobler
John G. Taylor

Logic and Planning
Susanne Biundo
Patrick Doherty
Henry Kautz
Paolo Traverso

Algebraic Methods in Logic
Miklos Ferenczi
Robin Hirsch
Idiko Sain

Non-monotonic Logics and Logics of Change
Jurgen Dix
Vladimir Lifschitz
Donald Nute
David Pearce

Logic and Learning
Luc de Raedt
John Lloyd
Steven Muggleton

Logic and Natural Language Processing
Wojciech Buszkowski
Hans Kamp
Marcus Kracht
Johanna Moore
Michael Moortgat
Manfred Pinkal
Hans Uszkoreit

Fuzzy Logic Uncertainty and Probability
Didier Dubois
Petr Hajek
Jeff Paris
Henri Prade
George Metcalfe
Jon Williamson

Scope and Submissions

This journal considers submission in all areas of pure and applied logic, including:

pure logical systems	dynamic logic
proof theory	quantum logic
constructive logic	algebraic logic
categorical logic	logic and cognition
modal and temporal logic	probabilistic logic
model theory	logic and networks
recursion theory	neuro-logical systems
type theory	complexity
nominal theory	argumentation theory
nonclassical logics	logic and computation
nonmonotonic logic	logic and language
numerical and uncertainty reasoning	logic engineering
logic and AI	knowledge-based systems
foundations of logic programming	automated reasoning
belief revision	knowledge representation
systems of knowledge and belief	logic in hardware and VLSI
logics and semantics of programming	natural language
specification and verification	concurrent computation
agent theory	planning
databases	

This journal will also consider papers on the application of logic in other subject areas: philosophy, cognitive science, physics etc. provided they have some formal content.

Submissions should be sent to Jane Spurr (jane.spurr@kcl.ac.uk) as a pdf file, preferably compiled in LaTeX using the IFCoLog class file.

Contents

ARTICLES

Formal Approaches to the Ontological Argument 1431
Ricardo Sousa Silvestre and Jean-Yves Béziau

A Brief Critical Introduction to the Ontological Argument and its Formalization: Anselm, Gaunilo, Descartes, Leibniz and Kant 1439
Ricardo Sousa Silvestre

A Mechanically Assisted Examination of Begging the Question in Anselm's Ontological Argument 1473
John Rushby

A Tractarian Resolution to the Ontological Argument 1497
Erik Thomsen

On Kant's Hidden Ambivalence Toward Existential Generalization in his Critique of the Ontological Argument 1515
Giovanni Mion

The Totality of Predicates and the Possibility of the Most Real Being 1523
Srećko Kovač

An Even More Leibnizian Version of Gödel's Ontological Argument 1553
Kordula Świętorzecka and Marcin Łyczak

A Case Study On Computational Hermeneutics: E. J. Lowe's Modal Ontological Argument **1567**
David Fuenmayor

Formal Approaches to the Ontological Argument

Ricardo Sousa Silvestre
Federal University of Campina Grande, Brasil.
ricardoss@ufcg.edu.br

Jean-Yves Béziau
Federal University of Rio de Janeiro, Brasil.
jyb@uni-log.org

Abstract

This paper presents the special issue on Formal Approaches to the Ontological Argument and briefly introduces the ontological argument from the standpoint of logic and philosophy of religion (more specifically the debate on the rationality of theistic belief).

Arguments for and against the existence of God have been proposed and subjected to logical analysis in different periods of the history of philosophy. In an important sense, they all deal with the rationality of theist belief. Providing a good argument for the conclusion that God does exist, or that it is highly probable that he exists, might be a pretty strong case for the thesis that belief in his existence is rational. Similarly, a good argument for the conclusion that God does not exist could be said to support the thesis that theistic belief is irrational.

A more basic approach than that would be to analyze the very concept of God. Can God create a stone so heavy that he cannot lift? If we say yes, then there is something God cannot do, namely to create such a stone; if we say no, there is also something he cannot do, namely to lift the stone. In either case he is not omnipotent. If really unsolvable, paradoxes like this (this is the paradox of the stone) show that the concept of God (who is, besides other things, omnipotent[1]) is incoherent or contradictory. Like the concept of a squared circle, it could never be

---

[1] For more on the concept of omnipotence and the paradox of the stone see [6].

instantiated. Theists must of course try refute claims like this; ideally, they must provide arguments showing that the concept of God is coherent or consistent.[2]

One of the most influential theist arguments (which is in fact a family of related arguments) in the history of philosophy is the ontological argument. First proposed by Anselm of Canterbury in the Eleventh Century, the ontological argument has been either analyzed or reformulated in the modern period by philosophers such as Descartes, Spinoza, Leibniz, Hume and Kant.[3] There has been a revival in the interest in the ontological argument in the Twentieth century; besides a growing literature on the topic, contemporary thinkers such Norman Malcolm [10], Charles Hartshorne [5], David Lewis [9], Alvin Plantinga [16] and Kurt Gödel [4] have either offered fresh views on the ontological argument or proposed new versions of it. It is by far the most commented argument for the existence of God — indeed, for the existence of anything — of the last half-century.

It is not difficult to see why this is so. To start with, the ontological argument is one of the most audacious arguments in the history of ideas. It is an *a priori* proof for the existence of God: from the mere concept of God, or from the mere definition of the word "God", it aims to arrive at the conclusion that God, as an ontological entity, exists in reality. Besides, it unities the two approaches to appraising the rationality of theistic belief mentioned above: the construction and analysis of arguments for and against the existence of God and the analysis of the concept of God.[4] In fact, the ontological argument is the most illustrious part of the most traditional and perhaps also the most comprehensive project of analyzing the concept of God: perfect being theology.

Perfect being theology is the endeavor of, from some definition of God as a maximally perfect being, derive conclusions about him, such as that he is unique, omniscient, omnipotent, morally perfect, omnipresent, eternal, impassible, simple and that he exists in reality (this is the ontological argument). Anselm was the first to do that; Descartes, Spinoza and Leibniz have also engaged in the same kind of project. Leibniz was the first to not take for granted that all perfections are compos-

---

[2]We are here using the terms "contradictory" and "consistent" as applied also to concepts. A concept C is consistent or coherent if and only if the set composed by "There is an object x which is C.", "The concept of C is defined ..." and whatever other sentence is needed to turn the definition into a complete one, is consistent. A non-consistent concept is called contradictory or non-coherent. Here is an example. The concept of squared cirque is contradictory, for the set {"There is an object x which is a squared circle.", "A squared circle is defined as a figure which, as a square, has four sides and, as a circle, has no sides.", "If a figure has no sides, then it is false that it has four sides."} is not consistent.

[3]See [19] for a brief historical introduction to the ontological argument

[4]For more on arguments for and against the existence of God see [11, chapters 4 to 7] and [22, 42 to 61]; for more on the analysis of the concept of God see chapters [11, 1 to 3] and [22, 25 to 41].

sible; attempting to fill what he took to be a shortcoming in Descartes' ontological arguments, he endeavored to show that all perfections can co-exist together in a single entity, or that it is possible that there is such a supremely perfect being, or still that the concept of God is not incoherent or contradictory. Leibniz's so-called ontological argument might therefore be seen as an argument for the coherence or consistency of the concept of God.[5]

Consistency is a logical concept; and arguments are the main object of study of logic. One might therefore justifiably think that logic as a field had played a strong role in the philosophical enquiry on God. That is not completely true. Unfortunately, the use of formal tools (which so distinctively characterize modern logic) in the construction and analysis of arguments for and against the existence of God and in the analysis of the concept of God is still an exception. One thing is to propose an argument and even analyze it (perhaps identifying premises and hidden presupposition and conclusion and seeing to what extent the former entails the latter) using ordinary language and common reasoning; other thing is to do that with the help of a logical language and a formal theory of inference, or to refer to pertinent results of modern logic. Otherwise stated, there is an important distinction between the use of logic as an indispensable component of any rational discourse and the use of tools and results of the field we call logic.

We are concerned here with this second, formal approach to the study arguments. Incidentally, the ontological argument also occupies a prominent place in this regard. It is perhaps the philosophical argument that received most formal treatment in the twentieth century. First, there have been in the past decades quite a good number of attempts to formally analyze several traditional versions of the ontological argument. Attempts to formally analyze the arguments attributed to Anselm, for instance, are abundant ([5, pp. 49–57], [1, 15, 7], [20, pp. 60–65], [12, 2]). Second, there have been many new formulations of the ontological argument directly embedded in formal frameworks.[6]

This special issue on formal approaches to the ontological argument contains both approaches, but in an unbalanced way. While one of the papers proposes a reformulation of Gödel's ontological argument, the other six deal mainly with existing ontological arguments and existing contributions to the debate on the ontological argument. Excepting for this introduction and the second paper, all articles which compose this special issue were delivered at the 2nd World Congress on Logic and Religion, which took place in Warsaw, Poland, on June 18–22, 2018.

The second paper, by Ricardo Silvestre (2018), tries to reach two goals. First,

---

[5]For more on perfect being theology see [14, 17] and [24].
[6]Gödel [4] and, to a lesser extent, [16] are instances of this.

it tries to function as an introduction to the ontological argument. As such, it complements this introductory paper, allowing readers not familiar with philosophical literature to have a better glimpse of the historical nuances related to the ontological argument. Secondly, it attempts at critically investigating the enterprise of formally analyzing philosophical arguments and, according to the author, contribute in a small degree to the debate on the role of formalization in philosophy. Silvestre approaches the issue from a Carnapian viewpoint: he sees the task of formalizing existing arguments as an explanatory endeavor, where the original argument is the *explicandum* and the formalized argument result of the analysis the *explicatum*. According to Carnap's theory of explication, the satisfactoriness of the *explicatum* can be assessed through four requirements; Silvestre refers to one in particular, the similarity requirement, adding a fifth one which is basically a development on the fruitfulness requirement.

The third paper, by John Rushby [18], deals with Anselm's argument. More precisely, it deals with a charge quite often made against ontological arguments in general and Anselm's argument in particular: that it is question begging. The way Rushby deals with the issue allows us to classify his work inside of what we might call computational philosophy: he uses a specific theorem prover named PVS to analyze several alternative formalizations of Anselm's Ontological Argument. By using a couple of different definitions of question begging, he concludes that all formalizations resort to some kind of question begging. Rushby's general thesis is that mechanized verification provides an effective and reliable technique to perform this kind of analysis.

The fourth paper, by Erik Thomsen [23], deals with Descartes' ontological argument. It however has a more general goal: to deal with what the author takes as the two major problems with ontological arguments: existential implications and term semantics. In order to exemplify them, he uses a specific reconstruction of Descartes' ontological argument and a less known one which makes blatant use of the rule of existential generalization. He then offers a resolution — in the sense of showing the fundamental mistakes that occur in an argument and how these mistakes reflect a foundational problem that lies at the heart of traditional logical views on existence and predication — for these two arguments by using a tractarian logic, that is to say, a logic which follows or is at least consistent with the principles laid out in Wittgenstein's *Tractatus*. These principles include a radical reinterpretation of the components of a proposition that defines logical subjects and functions/predicates in terms of sequenced computational processes instead of as references to general objects and properties.

The fifth paper, by Giovanni Mion [13], deals with Kant's critique against Descartes' argument. More specifically, it deals with a supposed problem in Kant's approach to existence in his critique of the ontological argument: If existence is not a predicate, but a quantifier, then a specific version of the ontological argument which uses the rule of existential generalization is valid. On the other hand, if existence is a real predicate, then the traditional Cartesian version of the ontological argument is sound. Mion's main goal is to provide a solution to this paradox. Besides, since he assumes that for Kant existence is governed by the rule of existential generalization, he also provides a proof for the following biconditional: existence is not a real predicate iff existential generalization is valid.

The sixth paper, by Srecko Kovac [8], deals with a very key issue in the ontological argument: the possibility of the most perfect or real being. It assesses Descartes' and Leibniz's ontological arguments; the verdict is that they fail because they do not distinguish between real and logical predicates (Descartes) or because they only show the logical possibility of such a perfect being, whereas the real possibility is what should be proved (Leibniz). Kovac then moves to Kant. He argues that Kant's doctrine of "transcendental ideal of pure reason" contains, in a rudimentary sense, a second-order axiomatic theory of reality (as a property of properties) and of the highest being, which he formalizes and which, he claims, anticipates Gödel's axiomatic proof for the possibility of a supreme being. The work is completed by offering such a proof for the possibility of the most real being.

The seventh paper, by Kordula Świętorzecka and Marcin Łyczak [21], deals with Leibniz and Gödel's ontological argument. More specifically, it proposes a modification of Gödel's ontological argument following what the authors call a Leibnizian onto-theology. The basic idea is to preserve the main structure of the Gödelian argument while taking some of Leibniz's ideas contained in some of his letters from 1676 and 1677 into account. First of all, Świętorzecka and Łyczak try to bring Gödel's concept of positiveness closer to the idea of a Leibnizian *perfectio*, which should not be understood via negations. Second, they analyse the concept of a necessary being in terms of a Leibnizian notion of demonstrability. To this end, they formulate an S4 version of Gödel's argument without using negative predicate terms. Finally, they sketch a model for their theory that allows, they argue, to express a few specific properties of the Leibnizian God.

The eighth and last paper, by David Fuenmayor and Christoph Benzmüller, is one more instance of what we called computational philosophy. It deals with a contemporary version of the ontological argument by analytic philosopher Edward Jonathan Lowe. The way Fuenmayor and Benzmüller build their formalization of Lowe's argument is very interesting. The idea is to work iteratively on the argument by temporarily fixing truth-values and inferential relations among its sentences, and then, after choosing a logic for formalization, working back and forth on the formalization of its axioms and theorems by making gradual adjustments while getting automatic feedback about the suitability of the decisions taken. They thus arrive at different variations or formalizations of the argument, each one, they claim, being an improvement on the previous one. They use a generic proof assistant called Isabelle, which uses a kind of higher-order logic dialect. Fuenmayor and Benzmüller call this method of formalizing arguments computation hermeneutics.

References

[1] R. Adams. The Logical Structure of Anselm's Arguments. *The Philosophical Review* 80: 28-54, 1971.

[2] G. Eder and E. Ramharte. Formal Reconstructions of St. Anselm's Ontological Argument. *Synthese* 192: 2791-2825. 2015.

[3] D. Fuenmayor and C. Benzmüller. A Case Study on Computational Hermeneutics: E. J. Lowe's Modal Ontological Argument. *The Journal of Applied Logics - IfCoLog Journal of Logics and their Applications* **: ***-****. 2018.

[4] K. Gödel. *Collected Works, vol. 3*. Oxford: Oxford University Press, 1995.

[5] C. Hartshorne. *The Logic of Perfection*. LaSalle, IL: Open Court, 1962.

[6] J. Hoffman and G. Rosenkrantz. Omnipotence. *The Stanford Encyclopedia of Philosophy (Winter 2017 Edition)*, (ed.) Edward N. Zalta (ed.), https://plato.stanford.edu/archives/win2017/entries/omnipotence/, 2017.

[7] G. Klima. Saint Anselm's Proof: A Problem of Reference, Intentional Identity and Mutual Understanding. In: Ghita Holmström-Hintikka (ed.), *Medieval Philosophy and Modern Times*, Dordrecht: Kluwer, 2000, p. 69-87.

[8] S. Kovac. The Totality of Predicates and the Possibility of the Most Real Being. *The Journal of Applied Logics - IfCoLog Journal of Logics and their Applications* **: ***-****. 2018.

[9] D. Lewis. Anselm and Actuality. *Noûs* 4: 175Ð88. 1970.

[10] N. Malcolm. Anselm's Ontological Arguments. *The Philosophical Review* 69: 41Ð62. 1960.

[11] W. Mann, ed. *The Blackwell Guide to the Philosophy of Religion*. Oxford: Blackwell Publishing, 2005.

[12] R. E. Maydole. The Ontological Argument. In: *The Blackwell Companion to Natural Theology*, (eds.) W. L. Craig and J. P. Moreland, Oxford: Blackwell, 2009, pp. 553Ð592.

[13] G. Mion. On Kant's hidden ambivalence toward existential generalization in his critique of the ontological argument. *The Journal of Applied Logics - IfCoLog Journal of Logics and their Applications* **: ***-****. 2018.

[14] T. Morris. Perfect Being Theology. *Noûs* 21: 19Ð30, 1987.

[15] P. Oppenheimer and E. Zalta. On the Logic of the Ontological Argument. In: *Philosophical Perspectives 5: The Philosophy of Religion*, (ed.) James Tomblin, Atascadero: Ridgview Press, 1991.

[16] A. Plantinga. *The Nature of Necessity*. Oxford: Oxford University Press, 1974.

[17] K. Rogers. *Perfect Being Theology*. Edinburgh, Edinburgh University Press, 2000.

[18] J. Rushby. A Mechanically Assisted Examination of Begging the Question in Anselm's Ontological Argument. *The Journal of Applied Logics - IfCoLog Journal of Logics and their Applications* **: ***-****. 2018.

[19] R. Silvestre. A Brief Critical Introduction to the Ontological Argument and its Formalization: Anselm, Gaunilo, Descartes, Leibniz and KantÓ. *The Journal of Applied Logics - IfCoLog Journal of Logics and their Applications* **: ***-****. 2018.

[20] J. Sobel. *Logic and Theism*. New York: Cambridge University Press, 2004.

[21] K. Świętorzecka and M. Łyczak. An Even More Leibnizian Version of Gödel's Ontological Argument. *The Journal of Applied Logics - IfCoLog Journal of Logics and their Applications* **: ***-****. 2018.

[22] C. Taliaferro, P. Draper and P. Quinn, eds. *A Companion to Philosophy of Religion*. Oxford: Blackwell Publishing, 2010.

[23] E. Thomsen. A Tractarian Resolution to the Ontological Argument. *The Journal of Applied Logics - IfCoLog Journal of Logics and their Applications* **: ***-****. 2018.

[24] M. Webb. Perfect Being Theology. In: *A Companion to Philosophy of Religion*, (eds.) C. Taliaferro, P. Draper and P. Quinn, Oxford: Blackwell Publishing, pp. 226-234, 2010.

A Brief Critical Introduction to the Ontological Argument and its Formalization: Anselm, Gaunilo, Descartes, Leibniz and Kant

Ricardo Sousa Silvestre
Federal University of Campina Grande, Brasil.
`ricardoss@ufcg.edu.br`

Abstract

The purpose of this paper is twofold. First, it aims at introducing the ontological argument through the analysis of five historical developments: Anselm's argument found in the second chapter of his Proslogion, Gaunilo's criticism of it, Descartes' version of the ontological argument found in his Meditations on First Philosophy, Leibniz's contribution to the debate on the ontological argument and his demonstration of the possibility of God, and Kant's famous criticisms against the (cartesian) ontological argument. Second, it intends to critically examine the enterprise of formally analyzing philosophical arguments and, as such, contribute in a small degree to the debate on the role of formalization in philosophy. My focus will be mainly on the drawbacks and limitations of such enterprise; as a guideline, I shall refer to a Carnapian, or Carnapian-like theory of argument analysis.

1 Introduction

The ontological argument is one of the most famous arguments (or family of arguments, to be more precise) in the history of philosophy. It was proposed in full-fledged form for the first time by Anselm of Canterbury, and either analyzed or reformulated by philosophers such as Descartes, Spinoza, Leibniz, Hume and Kant. Besides these classical approaches, so to speak, contemporary thinkers such as Norman Malcolm [22], Charles Hartshorne [11], David Lewis [21], Alvin Plantinga [29]) and Kurt Gödel [9] have either offered fresh views on the ontological argument or proposed new versions of it.

I would like to thank Giovanni Mion and Srecko Kovac for comments on a earlier draft of the paper.

The ontological argument is also perhaps the argument that has most attracted the attention of formal philosophers. Attempts to formally analyze the arguments attributed to Anselm, for instance, are abundant [11, pp. 49–57], [1, 26, 17, 24, 7] and [31, pp. 60–65]. Although there have been new formulations of the ontological argument directly embedded in formal frameworks,[1] the most common enterprise is still the formal analyses of traditional (and non-formal) versions of the ontological argument.

As far as formal analysis of existing philosophical arguments is concerned, some steps might be identified. First, there must be some sort of previous, informal analysis of the argument, meant to say, for example, what the premises and conclusion of the argument are, whether or not there are subsidiary arguments and hidden premises, etc. Second, there must be a formal language in which premises and conclusion are represented. Third, there might be an attempt to reconstruct the inferential steps of the original argument, possibly inside a specific theory of inference, be it proof theoretical or semantical or both. In a sense, the whole thing can be seen from the viewpoint of Carnap's project of conceptual explanation [5, pp. 1–18]. On one side, we have an argument, in general a prose text, whose relevant aspects — premises and conclusion, presuppositions, structure, etc. — are obscure and ambiguous. This would correspond to Carnap's notion of *explicandum*. On the other hand, we have the outcome of the analysis: a representation of the argument, possibly accompanied by a derivation, embedded in a formal framework, which is supposed to be a reconstruction, or to use Carnap's terminology, an explanation of the original argument. This is the *explicatum*.

Due to its exactness or formal feature, let us say, the *explicatum* is supposed not to have those obscure features of the *explicandum*. In particular, it must be evident in the *explicatum* the exact meaning of premises, conclusion and hidden presuppositions, the structure of the argument, and whether or not it is valid. The *explicatum* is also supposed to help in the evaluation of the reasonableness of the premises. This has to do with Carnap's second requirement: that the *explicatum* must be fruitful. Due to this, as well as to the very nature of formal reconstructions (Carnap would probably say their exactness) and the obscurity and incompleteness of informal arguments, the *explicatum* shall most probably have many features not shared by the original argument. However, this must not cause it to depart too much from the original argument, otherwise the former cannot be said to be an explanation of the latter. In Carnap's [5, p. 5] words, "the *explicatum* must be as close to or as similar with the *explicandum* as the latter's vagueness permits." To

---

[1] Gödel [9] and, to a lesser extent, Plantinga [29] are instances of this.

these three requirements — exactness, fruitfulness and similarity[2] — I will add a fourth one: that the *explicatum* should not be troublemaker, by which I mean that the *explicatum* or formal reconstruction should neither produce problems, confusing questions and unfruitful issues which are not already present in the *explicandum* nor obscure important and otherwise clear aspects of it.

The formal analysis of existing philosophical arguments can be categorized inside the umbrella of formalization in philosophy. As a methodology, the use of formal tools in philosophy has been the object of much debate in recent years [14, 10, 8]. Among other issues is the relation between formal philosophy and non-formal philosophy. Sven Hansson (2000) has rather dramatically put this as follows:

> Few issues in philosophical style and methodology are so controversial among philosophers as formalization. Some philosophers consider texts that make use of logical or mathematical notation as nonphilosophical and not worth reading, whereas others consider non-formal treatments as—at best—useful preparations for the real work to be done in a formal language. [...] This is unfortunate, since the value—or disvalue—of formalized methods is an important metaphilosophical issue that is worth systematic treatment. [...] It is urgently needed to revitalize formal philosophy and increase its interaction with non-formal philosophy. Technical developments should be focused on problems that have connections with philosophical issues.[3]

He correctly points out, although not that explicitly, that in order to revitalize formal philosophy and increase its interaction with non-formal philosophy, there must be a very clear understanding of the dangers and exaggerations of formalization [10, pp. 168–170].

The purpose of this paper is twofold. First, it aims at introducing the ontological argument through the analysis of five historical developments: Anselm's argument found in the second chapter of his *Proslogion*, Gaunilo's criticism of it, Descartes' version of the ontological argument found in his *Meditations on First Philosophy*, Leibniz's contribution to the debate on the ontological argument and his demonstration of the possibility of God, and Kant's famous criticisms against the (cartesian) ontological argument.

Second, it intends to critically examine the enterprise of formally analyzing philosophical arguments and, as such, contribute in a small degree to the debate on the role of formalization in philosophy. For this purpose, in my presentation of Anselm's

---

[2]There is a fourth requirement in Carnap's theory of conceptual explanation: simplicity [5, pp. 5–8].

[3]Hansson [10, pp. 162, 173].

argument and Gaunilo's criticism I shall refer to Robert Adam's (1971) pioneer work on the formalization of the ontological argument. Descartes' argument shall be introduced with the help of Howard Sobel's [31, pp. 31–40] analysis; as far as Leibniz's argument is concerned, I shall refer to Graham Oppy's [27, pp. 24–26] analysis, which, albeit not being a formal one, shall be useful as an instance of the first step in the task of formally analyzing an argument which I have mentioned above. My focus will be mainly on the drawbacks and limitations of these approaches as attempts to analyze existing philosophical arguments; as a guideline, I shall strongly refer to the Carnapian (or Carnapian-like) theory of argument analysis sketched above, specially its similarity and non-troublesome criteria.

The structure of the paper is as follows. In the next section I present Anselm's ontological argument, followed by Gaunilo's objection to it (Section 3)[4]. In Section 4, Descartes' version of the ontological argument is presented. In Section 5 I introduce Leibniz's contribution, which is followed by Kant's criticisms in Section 6. In Section 7 I lay down my concluding remarks about the enterprise of formally analyzing arguments.

2 Anselm

Although it is a consensus that a complete version of the ontological argument was first proposed by Anselm of Canterbury in his *Proslogion* (written between 1077 and 1078)[5], it is somehow controversial what the main argument is and where exactly in the text it is located.[6] Despite this, it is pretty safe to take the following extract from the second chapter of the Proslogion as describing the first, and surely the most famous, of Anselm's ontological arguments:

> (1) Well then, Lord, You who give understanding to faith, grant me that I may understand, as much as You see fit, that You exist as we believe You to exist, and that You are what we believe You to be. (2) Now we believe that You are something than which nothing greater can be thought. (3) Or can it be that a thing of such a nature does not exist, since "the Fool has said in his heart, there is no God?" (Psalms14, 1.1, and 53, l. 1.) (4) But surely, when this same Fool hears what I

---

[4]The content of Sections 2 and 3 has been partially taken from [32].

[5]The basic ideas of the *Proslogion* were anticipated in one of Anselm's earlier writings, the *Monologion*.

[6]While some authors ([6, 2]) believe that the major argument is found in the second chapter of the Proslogion, others ([22, 2, 29]) claim that the main argument is a modal one occurring in the third chapter. Still others [18] claim that the second and third chapter, and perhaps the entire work, comprise a single argument.

am talking about, namely, "something-than-which-nothing-greater-can-be-thought", he understands what he hears, and what he understands is in his mind (intellect, understanding), even if he does not understand that it actually exists. (5) For it is one thing for an object to exist in the mind, and another thing to understand that an object actually exists. (6) Thus, when a painter plans before hand what he is going to execute, he has (it) in his mind, but does not yet think that it actually exists because he has not yet executed it. (7) However, when he has actually painted it, then he both has it in his mind and understands that it exists because he has now made it. (8) Even the Fool, then, is forced to agree that something-than-which-nothing-greater-can-be-thought exists in the mind, since he understands this when he hears it, and whatever is understood is in the mind. (9) And surely that-than-which-a-greater-cannot-be-thought cannot exist in the mind alone. (10) For if it exists solely in the mind even, it can be thought to exist in reality also, which is greater. (11) If then that-than-which-a-greater-cannot-be-thought exists in the mind alone, this same that-than-which-a-greater-cannot-be-thought is that-than-which-a-greater-can-be-thought. (12) But this is obviously impossible. (13) Therefore there is absolutely no doubt that something-than-which-a-greater-cannot-be-thought exists both in the mind and in reality.[7]

Sentences (1) and (2) might be seen as an introduction to the argument. While (1) is a sort of opening statement, (2) is Anselm's famous definition of God: God is something than which nothing greater can be thought. (3) marks the proof style Anselm adopted: the reductio ad absurdum method; it states the reductio ad absurdum hypothesis, that is, the negation of what is supposed to be proved. Sentences (4) to (8) can be taken as a preliminary argument meant to prove a key premise of the argument: that something-than-which-nothing-greater-can-be-thought exists in the Fool's mind. (9) is an anticipation of the argument's conclusion: that God exists both in reality and in the understanding. (10) is the basic step of the argument: if this thing exists only in the mind, it can be thought to exist in reality also, and to exist in reality is greater. Sentence (11) states the consequence of what has been said so far: if this thing exists only in the mind, it will be at the same time that-than-which-a-greater-cannot-be-thought and that-than-which-a-greater-can-be-thought. But this, as sentence (12) says, is impossible. Therefore, the conclusion of the argument (13): that something-than-which-a-greater-cannot-be-thought exists both in the mind and in reality.

---

[7]Translation by M. J. Charlesworth [6].

In his pioneer work, Robert Adams [1, pp. 29–34] analyzes this argument as follows:

i There is, in the understanding at least, something than which nothing greater can be thought;

ii If it is even in the understanding alone, it can be thought to be in reality also;

iii which is greater;

iv There exists, therefore, ... both in the understanding and in reality, something than which a greater cannot be thought

As far as our numeration of Anselm's statements is concerned, (i) is (8), (ii) and (iii) are (10), and (iv) is (13). Adams still refers to a fourth premise which corresponds to the reductio ad absurdum hypothesis (3) and is used only at the time of reconstructing the derivation:

v There is no God.

A pertinent observation to be made about Adams's analysis concerns his choice of taking (8) as premise. As I have said, sentences from (4) to (8) can be taken very reasonably as a preliminary argument: while (4) is an anticipation of the conclusion and (8) is the conclusion, sentences (5) to (7) seem to be meant to support (8). That Adams skips this and takes (8) instead as premise is significant for a couple of reasons. First, although one could try to justify this move, the fact that Adams does not even mention it and simply ignores a good part of Anselm's original argument makes his analysis less faithful to it. Second, as an obvious consequence of that, Carnap's similarity criterion will probably not be satisfactorily met by Adams' reconstruction. Third, neglecting that Anselm himself tried to justify (8) has important consequences for evaluating the reasonableness of (Adam's reconstruction of) Anselm's argument. Some have argued that this premise is a very key one, and unless it is well justified, the argument as a whole might be accused of question-begging [30, pp. 37–52].

For the formalization proper, four predicates are used: $U(x)$, meaning that x exists in the understanding, $R(x)$, meaning that x exists in reality, $G(x, y)$, meaning that x is greater than y, and $Q(x, y)$, meaning that x is the magnitude of y. Besides them, Adams also uses a modal operator of possibility, represented here as $\Diamond$; $\Diamond \alpha$ means that it is possible that α,[8] which he takes as equivalent to *it can be thought that* α. Anselm's concept of a thing than which nothing greater can be though is represented as an abbreviation:

---

[8] Adams uses M instead of the symbol $\Diamond$.

$\phi(x, m) =_{\text{def}} Q(m, x) \land \neg \Diamond \exists y \exists n (G(n, m) \land Q(n, y))$

And here are the premises and conclusion of the argument:

I $\exists x \exists m (U(x) \land \phi(x, m))$

II $\forall x \forall m (U(x) \land \phi(x, m) \to \Diamond R(x))$

III $\forall x \forall m (\phi(x, m) \land \neg R(x) \to \neg \Diamond \neg (R(x) \to \exists n (G(n, m) \land Q(n, x))))$

IV $\exists x \exists m (U(x) \land R(x) \land \phi(x, m))$

$\phi(x, m)$ says that m is the magnitude of x and it is not possible that there is another thing, say y, whose magnitude n is greater than m. (I), (II) and (IV) are of easy understanding. (III) says that to every x and m, if x is God and m is his magnitude but he does not exist in reality, then it is not possible that the following proposition is false (that is to say, it is a necessary one): if x exists in reality, then its new magnitude, n, is greater than m. Adopting a counterfactual reading, (III) would mean the following: if God, whose magnitude is m, does not exist in reality, then would he exist in reality, his new magnitude, n, would be greater than m. The reductio ad absurdum hypothesis (v) is represented with the help of a constant, for it appears, one might argue, inside Anselm's talk-about-particulars discourse (see below):

V $\neg(a)$

Premise (III) — and the original sentence (10) in the argument — incorporates one of the most controversial issues in Anselm's argument, namely the doctrine that existence is something which 'produces' greatness:

(G) It is greater to exist in reality as well than to exist merely in the understanding.

In its turn, (G) might be understood in at least three different ways [23, pp. 90–91]:

(G1) Anything that exists both in reality and in the understanding is greater than anything that exists in the understanding alone.

(G2) Anything that exists both in reality and in the understanding is greater than the otherwise same kind of thing that exists in the understanding alone.

(G3) Anything that exists both in the understanding and in reality is greater than the otherwise exact same thing, if that thing exists merely in the understanding.

Adams picks (G3) as the correct or more suitable interpretation of (G). However, even considering its attempt to be as precise as possible, (G3) is still ambiguous with respect to one thing: are these two things we are comparing exactly the same object, or two objects which differ in one aspect only (existence)? Adams representation leaves no doubt: we are comparing the very and same object, the one referred to by variable x.

For the derivation, the following inference rules are used:[9]

M1. $\neg\Diamond\neg(\alpha \to \beta), \Diamond\alpha \vdash \Diamond\beta$

M2. $\exists x \Diamond \alpha(x) \vdash \Diamond \exists x \alpha(x)$

C1. $\exists x \alpha(x) \vdash \alpha(x/t)$

C2. $\forall x \alpha(x) \vdash \alpha(x/t)$

C3. $\alpha \wedge \beta, \beta \wedge \varphi \to \lambda \vdash \varphi \to \lambda$

C4. $\alpha(t) \vdash \exists x \alpha(t/x)$

C5. $\alpha \wedge \beta, \varphi \vdash \beta \wedge \varphi$

C6. If $\Gamma, \alpha \vdash \beta$ then $\Gamma \vdash \alpha \to \beta$

C7. $\neg\alpha \to \beta \wedge \neg\beta \vdash \alpha$

C8. $\alpha \wedge \beta, \varphi \vdash \alpha \wedge \varphi \wedge \beta$

MP. $\alpha, \alpha \to \beta \vdash \beta$

And here is the derivation:

1. $\exists x \exists m (U(x) \wedge \phi(x,m))$ Pr. (I)

2. $\forall x \forall m (U(x) \wedge \phi(x,m) \to \Diamond R(x))$ Pr.(II)

3. $\forall x \forall m (\phi(x,m) \wedge \neg R(x) \to \neg\Diamond\neg(R(x) \to \exists n (G(n,m) \wedge Q(n,x))))$ Pr. (III)

---

[9]Adams bases his formal treatment on Quine's *Methods of Logic*; I have here adopted a more standard notation. Some of these rules shall be used also in the coming sections. Due to their elementariness, I shall not bother neither to justify nor to prove them. Neither shall I take into consideration the provisos that some rules such as C1 and C2 are supposed to have, which depend on the particularities of the axiomatization at hand. If you wish, you could say that I am using a kind of a semi-formal approach here.

4. $U(a) \wedge \phi(a,b)$	C1 (2x) 1[10]
5. $U(a) \wedge \phi(a,b) \rightarrow \Diamond R(a)$	C2 (2x) 2
6. $\Diamond R(a)$	MP 5,4
7. $\phi(a,b) \wedge \neg R(a) \rightarrow \neg \Diamond \neg (R(a) \rightarrow \exists n(Gn,b) \wedge Q(n,a)))$	C2 (2x) 3
8. $\neg R(a) \rightarrow \neg \Diamond \neg (R(a) \rightarrow \exists n(G(n,b) \wedge Q(n,a)))$	C3 4, 7
*9. $\neg R(a)$	Pr. (V)
*10. $\neg \Diamond \neg (R(a) \rightarrow \exists n(G(n,b) \wedge Q(n,a)))$	MP 8,9
*11. $\Diamond \exists n(G(n,b) \wedge Q(n,a))$	M1 6,10
*12. $\exists y \Diamond \exists n(G(n,b) \wedge Q(n,y))$	C4 11
*13. $\Diamond \exists y \exists n(G(n,b) \wedge Q(n,y))$	M2 12
*14. $U(a) \wedge Q(b,a) \wedge \neg \Diamond \exists y \exists n(G(n,b) \wedge Q(n,y))$	4[11]
*15. $\Diamond \exists y \exists n(G(n,b) \wedge Q(n,y)) \wedge \neg \Diamond \exists y \exists x(G(n,b) \wedge Q(n,y))$	C5 13,14
16. $\neg R(a) \rightarrow \Diamond \exists y \exists n(G(n,b) \wedge Q(n,y)) \wedge \neg \Diamond \exists y \exists x(G(n,b) \wedge Q(n,y))$	C6 9,15
17. $R(a)$	C7 16
18. $U(a) \wedge R(a) \wedge \phi(a,b)$	C8 4,17
19. $\exists x \exists m(U(x) \wedge R(x) \wedge \phi(x,m))$	C4 (2x) 18

A couple of things have to be said about this reconstruction of Anselm's argument. First, it is exactly this: a reconstruction. At most, it might be taken as revealing the logic beyond Anselm's argument or unclosing all otherwise hidden logical steps needed to turn Anselm's argument into a valid one. Trivially Anselm's argument does not have this structure; at no point of the text do we find evidence for most of the steps and inference rules that Adams uses.

Despite of this, and this is the second point, Adams correctly represents two important structural features of Anselm's argument. First, starting from step 9, it uses the *reductio ad absurdum* method found in the original argument (it ends at

---
[10] Here "C1 (2x) 1" means that this step is justified by applying two times rule C1 to formula of step 1.

[11] Here is the unabbreviated form of 4.

15). Second, Anselm's original argument switches back and forth from a universal discourse to talk about particulars. From (4) to (8) he speaks about something than which nothing greater can be thought; however, from (9) to (12) he changes his discourse and starts speaking about that than which a greater cannot be thought; then, in (13), he goes back to talk about something than which nothing greater can be thought. Adams correctly represents this movement.[12]

Third, about Adams' use of the operator $\Diamond$, sure it is an interesting way to represent the expression "it can be thought that". However, it is significant that Anselm's original argument does not use any kind of modal construction. We might therefore once more bring into scene Carnap's similarity criterion. Moreover, from a logical point of view, taking "it can be thought that" to be equivalent to "it is possible that" has some worrisome consequences. Trivially, the correctness of his reconstruction depends on the validity of the modal inferences he uses. That they are valid when interpreting $\Diamond$ as "it is possible that" is not a big issue. But how about Adams' interpretation? Is the validity of these modal inferences automatically transferred when one interprets $\Diamond$ as "it can be thought that"? It is somehow *ad hoc* to arbitrarily assume that this question can be answered with a "yes".

3 Gaunilo

The very first objection to Anselm's argument[13] was given by one of his contemporaries, the Marmoutier monk Gaunilo, in a pamphlet entitled "On Behalf of the Fool". Here are Gaunilo's words:

> Consider this example: Certain people say that somewhere in the ocean there is a "Lost Island" [...] which is more abundantly filled with inestimable riches and delights than the Isles of the Blessed. [...] Suppose that one was to go on to say: You cannot doubt that this island, the most perfect of all lands, actually exists somewhere in reality, because it

---

[12]Using C1 and C2, he switches, in steps 4, 5 and 7, from a universal discourse to discourse about particulars (in the case, individuals *a* and *b*). Similarly to Anselm's original argument, all crucial *reductio ad absurdum* steps are done inside this particular discourse framework. Then, when he has proved that *a* exists in reality, he goes back in step 19, thought C4, to the universal type of discourse.

[13]Anselm's argument has been attacked on several different grounds. It might be objected, for instance, that the concept of greatness used by Anselm unjustifiably presupposes the existence of a maximum. How about if the order relation involved in such a concept is alike to the order of natural numbers, that is to say, how about if for every being we can think of, it is always possible to think of something greater than it? For the sake of space, I shall here mention only the objections related to the historical development I am following.

undoubtedly stands in relation to your understanding. Since it is most excellent, not simply to stand in relation to the understanding, but to be in reality as well, therefore this island must necessarily be in reality. [...] If, I repeat, someone should wish by this argument to demonstrate to me that this island truly exists and is no longer to be doubted, I would think he were joking.[14]

Gaunilo's idea was to provide an argument which parallels Anselm's reasoning but which has an absurd conclusion. In order to reject the absurd conclusion that there exists such a lost perfect island, one has of course to reject the whole argument as invalid, even if she is unable to point out exactly what the defective steps in the argument are. But since the argument shares the same structure, so it is believed, than Anselm's argument, one is forced to also reject the latter argument along with its conclusion that there is something than which a greater cannot be thought.

In order to formalize Gaunilo's counter-argument, Adams [1, pp. 34–40] uses the following predicates: $I(x)$, meaning that x is an island, $L(x)$, meaning that x is a land or country, and $P(x)$, meaning that x has the profitable and delightful features attributed by legend to the lost island. The premises and conclusion of the argument, already formalized, are as follows:

(I) $\exists x(U(x) \land I(x) \land P(x) \land \neg \exists y(L(y) \land G(y,x)))$

(II) $\exists x(L(x) \land R(x))$

(III) $\forall x \forall y(L(x) \land R(x) \land I(y) \land \neg R(y) \to G(x,y))$

(IV) $\exists x(U(x) \land R(x) \land I(x) \land P(x) \land \neg \exists y(L(y) \land G(y,x)))$

(I) means that there is an individual x which exists in the understanding, is an island, has the profitable and delightful features attributed by legend to the lost island and, besides, there is no land greater than it. (II) says that there exists a real land. (III) says that any real land is greater than any island which does not exist in reality. The conclusion (IV) says that there exists such an island, both in the understanding and in reality, and that there is no greater land. Here is the *reductio ad absurdum* hypothesis:

(V) $\neg R(b)$

And here is the derivation:

1. $\exists x(U(x) \land I(x) \land P(x) \land \neg \exists y(Ly \land G(y,x)))$ \hfill Pr. (I)

---
[14]Hick and McGill [13, pp. 22-23].

2. $\exists x(L(x) \land R(x))$	Pr. (II)
3. $\forall x \forall y(L(x) \land R(x) \land I(y) \land \neg R(y) \to G(x,y))$	Pr (III)
4. $U(b) \land I(b) \land P(b) \land \neg\exists y(L(y) \land G(y,b)$	C1 1
5. $L(a) \land R(a)$	C1 2
6. $L(a) \land R(a) \land I(b) \land \neg R(b) \to G(a,b)$	C2 (2x) 3
7. $\neg R(b) \to G(a,b)$	C9 4,5,6
*8. $\neg R(b)$	Pr. (V)
*9. $G(a,b)$	MP 8, 7
*10. $L(a) \land G(a,b)$	C5 5,9
*11. $\exists y(L(y) \land G(y,b))$	C4 10
*12. $\exists y(L(y) \land G(y,b)) \land \neg\exists y(L(y) \land G(y,b)$	C5 4,11
13. $\neg R(b) \to \exists y(L(y) \land G(y,b)) \land \neg\exists y(L(y) \land G(y,b)$	C6 8, 12
14. $R(b)$	C7 13
15. $U(b) \land R(b) \land I(b) \land P(b) \land \neg\exists y(L(y) \land G(y,b))$	C8 4,14
16. $\exists x(U(x) \land R(x) \land I(x) \land P(x) \land \neg\exists y(L(y) \land G(y,x)))$	C4 15

where C9 is the following additional rule of inference:

C9. $\sigma_1 \land \beta \land \sigma_2, \alpha, \alpha \land \beta \land \lambda \to \varphi \vdash \lambda \to \varphi$

This is a valid argument. As far as Anselm's argument is concerned, despite the similarities (both proofs use the *reductio ad absurdum* method and the universal-to-particular-to-universal movement), it is pretty clear that both arguments have a quite different structure. In fact, the structure departure starts from the logical form of the premises: whereas Anselm spoke of a being whose greatness could not possibly be surpassed, Gaunilo speaks only of an island to which no country is superior.

Given this, it seems that Gaunilo's argument fails as a counter-argument to Anselm's. As I have said, a counter-argument in this sense is an argument that shares the same logical structure than the target argument, has true or reasonable premises and an absurd or patently false conclusion. But according to Adams' reconstructions both arguments have a quite different structure, which might allow us to conclude that, contrary to first appearances, Gaunilo did not succeed in refuting Anselm's argument.

This conclusion of course depends on the claim that Adams' formalization is a faithful and correct reconstruction of both Anselm's and Gaunilo's arguments. This of course is far from being trivial. As I have pointed out, Adams' reconstruction of Anselm's argument might be charged of departing too much from Anselm's original formulation, and therefore not satisfying Carnap's criterion of similarity to the *explanandum*. Besides, there are reconstructions according to which both arguments seem to share the same formal structure [27, pp. 17–18].

4 Descartes

Although writings such as the *Discourse on the Method* (1637) and *The Principles of Philosophy* (1644) discuss *a priori* arguments for the existence of God, Descartes' most referred version of the ontological argument appears in the fifth chapter of his *Meditations on First Philosophy*, first published in 1641. It is however controversial where exactly in the Fifth Meditation the argument (or arguments) is and how it shall be reconstructed. I shall take the following passage as the best representative of Descartes' formulation of his ontological argument:

> (1) [...] although it is not necessary that I should at any time entertain the notion of God, nevertheless whenever it happens that I think of a first and a sovereign Being, and, so to speak, derive the idea of Him from the storehouse of my mind, it is necessary that I should attribute to Him every sort of perfection, although I do not get so far as to enumerate them all, or to apply my mind to each one in particular. And this necessity suffices to make me conclude, (2) after having recognized that existence is a perfection, that (3) this first and sovereign Being really exists.[15]

Descartes' argument turns out to be a very simple one: the first premise (1) states that the idea or concept of God includes all perfections; from that, along with the second premise — that (2) existence is a perfection — we conclude that (3) God exists.[16]

In explaining the rationale behind this argument, it might be useful to refer to something which comes a little before this passage in the Fifth Meditation:

> But now, if just because I can draw the idea of something from my thought, it follows that all which I know clearly and distinctly as per-

---
[15]Translation by Elizabeth S. Haldane and G. R. T. Ross [19, p. 182].
[16]For alternative reconstructions of Descartes' ontological argument, see [27, pp. 20–24] and [25].

> taining to this object does really belong to it, may I not derive from this an argument demonstrating the existence of God? It is certain that I no less find the idea of God, that is to say, the idea of a supremely perfect Being, in me, than that of any figure or number whatever it is; and I do not know any less clearly and distinctly that an [actual and] eternal existence pertains to this nature than I know that all that which I am able to demonstrate of some figure or number truly pertains to the nature of this figure or number, and therefore, although all that I concluded in the preceding Meditations were found to be false, the existence of God would pass with me as at least as certain as I have ever held the truths of mathematics (which concern only numbers and figures) to be.[17]

Here Descartes invokes one of his key epistemological rules — that whatever I clearly and distinctly perceive to be contained in the idea of something is true of that thing — to introduce the possibility of an argument for the existence of God. In the same way that we arrive at basic truths about the nature of a figure or number, we might arrive at the conclusion that God exists simply by apprehending clearly and distinctively that existence pertains to the nature of such a supremely perfect being.

Although some have taken this passage as part of the argument itself [27, p. 21], it might be seen as providing a justification for the premises of the argument, first by offering the rule of clarity and distinctiveness as the epistemological support for the two premises, and second by giving a definition or explanation for the concept of God — a supremely perfect Being — which would render the first premise

(i) A supremely perfect being has every perfection.

quasi-tautological. Using this definition, the other premise and conclusion would be written as follows:

(ii) Existence is a perfection.

(iii) A supremely perfect being does exist.

This is the beginning of Sobel's reconstruction of Descartes argument [31, pp. 31–40].[18] Sobel points out that there is an ambiguity in (iii), which might be read either

---

[17]Translation by Elizabeth S. Haldane and G. R. T. Ross [19, pp. 180–181].

[18]Due to a passage in the Fifth Meditation where Descartes says that "it is no less repugnant to think of a God (that is, a supremely perfect being) lacking existence (that is, lacking some perfection), than it is to think of a mountain lacking a valley", Sobel takes him to be using a *reductio ad absurdum* proof style, adding then a fourth premise to the argument: A supremely perfect being does *not* exist. Even though one might argue against this analysis of Descartes' reasoning, it shall not interfere in my analysis of Sobel's contribution, for in order to get the contradiction one has to go through a direct proof, as Sobel does, and conclude that a supremely perfect being *does* exist.

as "Any supremely perfect being exists" or "At least one supremely perfect being exists". After investigating the consequences of reconstructing the argument using the first reading, he (correctly) picks the second one as the most accurate analysis of Descartes' argument.

For the formalization, Sobel uses three predicates — S, P and G — and a constant e. $S(x)$ means that x is a supremely perfect being, $P(x)$ that x is a perfection and $H(x, y)$ that x has property y; e means the property of existence. The argument is represented as follows:

(I) $\forall x(S(x) \to \forall y((P(y) \to H(x, y))))$

(II) $P(e)$

(III) $\exists x(S(x) \land H(x, e))$

The conclusion of Sobel's analysis is that the argument is invalid: trivially, from the two premises we cannot arrive at (III); at most we reach at the conclusion that $\forall x(Sx \to H(x, e))$. But here Sobel's analysis was extremely uncharitable to Descartes, to say the least. When a very key aspect of Descartes' argument is considered, we see that Sobel's analysis is faulty — it does not satisfactorily meet with the similarity criterion — and the argument straightforwardly valid.

Here is the key feature of Descartes' argument that Sobel misses: If some object has the property of existence, it obviously must exist. In other words, there must be some kind of link between the property of existence and existence itself. From the perspective of the formalism Sobel uses, this means that there must be a connection between constant e and the existential quantifier $\exists$ (recall that they represent the very same notion, namely existence in reality). This might be expressed as follows:

(IV) $\forall x(H(x, e) \to \exists x H(x, e))$

This unfortunately does not solve the issue. First of all, even with this extra axiom we cannot derive (III): in order to conclude $\exists x H(x, e)$ we should have $H(d, e)$ for some object d, which we cannot, for if we had $H(d, e)$, we would have the hard part of the conclusion — $\exists x H(x, e)$ — and the argument would be patently circular. Second, and this is trivially related to the first point, $H(d, e) \to \exists x H(x, e)$ is an instance of the rule of existential generalization. (IV) is tautological and, as such, adds nothing to our set of premises. Since classical first order logic requires each singular term to denote an object in the domain of quantification, which is usually understood as the set of existing objects, it is vacuous to say of an object that if it has some property it exists: in order to have any property, it must already exist.

This inability to properly represent what seems to be a very key presupposition of Descartes' argument reveals that Sobel's approach is misguided and his reconstruction a troublemaker one. But what if we take seriously Descartes' claim, found in both extracts of the Meditations shown above, that the idea of God is in our minds? Is not Descartes presupposing here a kind of existence pretty much alike to Anselm's notion of existence in the understanding? It seems to me that textual evidence suggests a positive answer to these questions.

What follows is an attempt to consider these ruminations and fix the issue still inside the basic logical framework which Sobel uses, that is to say, first-order classical logic.[19] Taking the existential quantifier (and consequently the universal quantification) to refer to this Anselmian-like notion of existence — in cartesian terms, $\exists x A(x)$ would mean that I can draw from my thought the idea of some x which has property A — and predicate H and constant e to the notion of existence in reality — $H(x, e)$ means that x exists in reality — the argument could be rewritten as follows:

(I*) $\exists x S(x)$

(I) $\forall x(S(x) \rightarrow \forall y(P(y) \rightarrow H(x,y)))$

(II) $P(e)$

(III) $\exists x(S(x) \wedge H(x,e))$

(I*) and (I) both represent premise (i). In the same way that there is an ambiguity in (iii), there is also an ambiguity in (i): it can be read as "Any supremely perfect being has every perfection" or "At least one supremely perfect being exists and it has every perfection." While the first reading is the one (and only one) that Sobel takes into account (to characterize the concept of a supreme being), the second one encapsulates the presupposition that a supreme being (or the idea of a supreme being, to be more precise) exists in our minds. Instead then of preferring one reading over the other, I take both of them into account; my reading of (i) in terms of (I*) and (I) is a compromise between these two interpretations.

The proof of the argument validity is straightforward:

1. $\exists x S(x)$ \hfill Pr. (I*)

2. $\forall x(S(x) \rightarrow \forall y(P(y) \rightarrow H(x,y)))$ \hfill Pr. (I)

3. $P(e)$ \hfill Pr. (II)

---

[19] This is why I have left out free logic, which in this context could be a better representational tool than classical first order logic.

4. $S(d)$	C1 1
5. $S(d) \to \forall y(P(y) \to H(d,y))$	C2 2
6. $\forall y(P(y) \to H(d,y))$	MP 4,5
7. $P(e) \to H(d,e)$	C2 6
8. $H(d,e)$	MP 3,7
9. $S(d) \land H(d,e)$	C10 4,8
10. $\exists x(S(x) \land H(x,e))$	C4 9

In addition to the rules introduced in Section 2 in the context of Adam's reconstruction of Anselm's argument, an additional one has been used here:

C10. $\alpha, \beta \vdash \alpha \land \beta$

A more elegant second order version of this reconstruction of Descartes' argument would be as follows:

(*I) $\exists x S(x)$

(I) $\forall x(S(x) \to \forall Y(\mathrm{P}(Y) \to Y(x)))$

(II) $\mathrm{P}(E)$

(III) $\exists x(S(x) \land E(x))$

where $P(Y)$ is a second order predicate meaning that Y is a perfection and $E(x)$ is a first order predicate meaning that x exists in reality. The derivation then would be rewritten as follows:

1. $\exists x S(x)$	Pr. (I*)
2. $\forall x(S(x) \to \forall Y(\mathrm{P}(Y) \to Y(x)))$	Pr. (I)
3. $\mathrm{P}(E)$	Pr. (II)
4. $S(d)$	C1 1
5. $S(d) \to \forall Y(\mathrm{P}(Y) \to Y(d))$	C2 2
6. $\forall Y(\mathrm{P}(Y) \to Y(d))$	MP 4,5

7. $P(E) \to E(d)$ C2² 6

8. $E(d)$ MP 3,7

9. $S(d) \wedge E(d)$ C10 4,8

10. $\exists x(S(x) \wedge E(x))$ C4 9

, where C2² is second order version of C2.

5 Leibniz

Leibniz wrote about ontological arguments in many of his works, including *Monadology* (1714), *Theodicy* (1710), and *New Essays Concerning Human Understanding* (completed in 1704). Although he presented at least three different ontological proofs for the existence of God,[20] his most important contribution to the history of ontological arguments is his attempt to demonstrate the coherence of the concept of God. If this is not done, he argued, all ontological arguments are irremediably defective. In the "Meditations on Knowledge, Truth, and Ideas" of 1684, for instance, he analyzes what he calls the "old argument for the existence of God" as follows:

> The argument goes like this: Whatever follows from the idea or definition of a thing can be predicated of the thing. God is by definition the most perfect being, or the being nothing greater than which can be thought. Now, the idea of the most perfect being includes ideas of all perfections, and amongst these perfections is existence. So existence follows from the idea of God. Therefore [...] God exists. But this argument shows only that if God is possible then it follows that he exists. For we can't safely draw conclusions from definitions unless we know first that they are real definitions, that is, that they don't include any contradictions. If a definition does harbour a contradiction, we can infer contradictory conclusions from it, which is absurd.[21]

Leibniz is here charging the ontological argument of being incomplete. He is concerned that the concept of a being who possesses all perfections might not be consistent, that is to say, that two perfections *A* and *B*, say, are such that there cannot be an individual that possess *A* and *B* at the same time. In order for the concept of God as defined by Anselm and Descartes to be a "real definition", one has to show first

---

[20] Blumenfeld [4] tries to how that the three proofs are equivalent, something in which Leibniz himself believed, he claims.

[21] Translation by Jonathan Bennet [3].

that all perfections or 'greatness producers' are compossible, that is to say, that it is possible for all of them to be instantiated by one and the same individual. Unless this is shown, all the ontological arguments have reached is the following conclusion: if God is possible, then he exists. As far as Anselm's and Descartes's arguments are concerned, this implies that the reasonableness of

(I) $\exists x \exists m (U(x) \wedge \phi(x, m))$

and

(I*) $\exists x S(x)$

depend, respectively, on the truth of $\Diamond \exists x \exists m \phi(x, m)$ and $\Diamond \exists x S(x)$.

In order to fill in this gap, Leibniz presents the following argument, found in his *New Essays concerning Human Understanding*:

> I call every simple quality which is positive and absolute, or expresses whatever it expresses without any limits, a perfection. But a quality of this sort, because it is simple, is therefore irresolvable or indefinable, for otherwise, either it will not be a simple quality but an aggregate of many, or, if it is one, it will be circumscribed by limits and so be known through negations of further progress contrary to the hypothesis, for a purely positive quality was assumed. From these considerations it is not difficult to show that all perfections are compatible with each other or can exist in the same subject. For let the proposition be of this kind:
>
> ***A*** and ***B*** are incompatible
>
> (for understanding by ***A*** and ***B*** two simple forms of this kind or perfections, and it is the same if more are assumed like them), it is evident that it cannot be demonstrated without the resolution of the terms ***A*** and ***B***, of each or both; for otherwise their nature would not enter into the ratiocination and the incompatibility could be demonstrated as well from any others as from themselves. But now (by hypothesis) they are irresolvable. Therefore this proposition cannot be demonstrated from these forms.[22]

This is the extract of Leibniz's work that Oppy analyzes. He informally reconstructs the argument as follows [27, p. 23]:

i. By definition, a perfection is a simple quality that is positive and absolute. (Definition)

---

[22]Translation by Alfred Langley [20, pp. 714–715].

ii. A simple quality that is positive and absolute is irresolvable or indefinable. (Premise — capable of further defense)

iii. *A* and *B* are perfections whose incompatibility can be demonstrated. (Hypothesis for *reductio*)

iv. In order to demonstrate the incompatibility of *A* and *B*, *A* and *B* must be resolved. (Premise)

v. Neither *A* nor *B* can be resolved. (From ii)

vi. (Hence) It cannot be demonstrated that *A* and *B* are incompatible. (From iii, iv, v by *reductio*)

Here is an attempt to turn this into a more formal and detailed account. Let ■ be a modal operator meant to represent the notion of demonstrability so that ■α means that α is demonstrable and $S(Y), O(Y), A(Y)$ and $R(Y)$ four second order predicates meaning, respectively, that Y is simple, Y is positive, Y is absolute and Y is resolvable. Besides, there are the following two definitions:

$$C(X, Y) =_{def} \Diamond \exists z(X(z) \wedge Y(z))$$
$$P(Y) =_{def} S(Y) \wedge O(Y) \wedge A(Y)$$

$C(X, Y)$ means that X and Y are two compossible (or compatible, in Oppy's terminology) properties and $P(Y)$ that Y is a perfection. The premises of the argument would then be represented as follows:

(I) $\forall Y(P(Y) \to \neg R(Y))$

(II) $P(A)$

(III) $P(B)$

(IV) $\forall X \forall Y(\blacksquare \neg C(X, Y) \to R(X) \wedge R(Y))$

(V) $\blacksquare \neg C(A, B)$

(I) is premise (ii), (II) and (III) are (partially) (iii), (IV) is the universal form of (iv) and (V) is the hypothesis for *absurdum*, which is also contained in (iii). The conclusion is obviously

(VI) $\neg \blacksquare \neg C(A, B)$

As for the derivation, the following additional rules shall be used:

1458

C7*. $\alpha \to \beta \wedge \neg \beta \vdash \neg \alpha$

C11. $\neg \alpha \vdash \neg(\alpha \wedge \beta)$

C12. $\alpha, \beta \vdash \alpha \wedge \beta$

And here is the reconstruction of the derivation:

1. $\forall Y(P(Y) \to \neg R(Y)$		Pr. (I)
2. $P(A)$		Pr. (II)
3. $P(A) \to \neg R(A)$		$C2^2$ 1
4. $\neg R(A)$		MP 2,3
5. $P(B)$		Pr. (III)
6. $P(B) \to \neg R(B)$		$C2^2$ 1
7. $\neg R(B)$		MP 5,6
8. $\forall X \forall Y(\blacksquare \neg C(X,Y) \to R(X) \wedge R(Y))$		Pr. (IV)
9. $\forall Y(\blacksquare \neg(C(A,Y) \to R(A) \wedge R(Y))$		$C2^2$ 8
10. $\blacksquare \neg C(A,B) \to R(A) \wedge R(B)$		$C2^2$ 9
*11. $\blacksquare \neg C(A,B)$		Pr. (V)
*12. $R(A) \wedge R(B)$		MP 11,10
*13. $\neg(R(A) \wedge R(B))$		C11 4
*14. $(R(A) \wedge R(B)) \wedge \neg(R(A) \wedge R(B))$		C12 12,13
15. $\blacksquare \neg C(A,B) \to (R(A) \wedge R(B)) \wedge \neg(R(A) \wedge R(B))$		C6 11,14
16. $\neg \blacksquare \neg C(A,B)$		C7* 15

Oppy offers two criticisms against Leibniz's attempt to fix the ontological arguments. First, he correctly points out that showing that it cannot be demonstrated that A and B are incompatible is quite different from showing that *A* and *B* are compatible, what makes it obvious that the argument failed to reach the required conclusion [27, pp. 25–26].

Two observations are in order here. First of all, the conclusion that $\neg\blacksquare\neg C(A, B)$ might be quite relevant to the debate about the ontological argument. The way Leibniz puts the whole thing, saying that all the ontological arguments show is that if God is possible then it follows that he exists, implies that the ontological defender is the one who has to bear the burden of the proof. But it might also be argued, as Leibniz himself did, that "there is always a presumption on the side of possibility; that is to say, everything is held to be possible until its impossibility is proved".[23] In other words, possibility claims are blameless until the contrary is proven. The burden of the proof then, in this case the proof that the concept of God is indeed incoherent, is on the critic of the ontological argument. But if this is correct, then showing that the concept of God cannot be proved to be incoherent has the relevant consequence that the critic's movement to refute the argument is hopeless.[24]

Second, Oppy oddly overlooks what Leibniz writes right after the quotation he analyzes, where he clearly does offer an argument for the conclusion that A and B are compossible:

> But it might certainly be demonstrated by these if it were true, because it is not true *per se*, for all propositions necessarily true are either demonstrable or known per se. Therefore, this proposition is not necessarily true. Or if it is not necessary that A and B exist in the same subject, they cannot therefore exist in the same subject, and since the reasoning is the same as regards any other assumed qualities of this kind, therefore all perfections are compatible.[25]

Complementing Oppy's analysis then, we would have the two additional premises:

vii. A proposition is necessary only it is true (or known) *per se* or demonstrable.

viii. That A and B are incompatible is not true (or known) *per se*,

which might be formalized as follows:

---

[23] Quoted from Blumenfeld [4, p. 357]. Blumenfeld defends that there is an argument in Leibniz, which he calls Leibniz's fallback position, that, in the absence of proof, one ought to assume that God is possible.

[24] This can also be seen from another angle. From a dialectical point of view, there are two movements one can make to criticize a valid argument. The first is trying to show that one of its premises is false or is known to be false; if successful, this movement would be enough to claim that the argument has been refuted. The second, and more modest movement, is to question the truth of one of the premises on the grounds that the defender was unable to provide strong support or evidence for it or did not prove some non-obvious presupposition on which it depends; it puts the burden of the proof on the defender's shoulder, claiming the argument to be incomplete, so to speak, but does not serve as a final word on the correctness of the argument.

[25] Translation by Alfred Langley [20, p. 715].

VII. $\Box \alpha \to \blacksquare \alpha \vee \circ \alpha$

VIII. $\neg \circ \neg C(A, B)$,

where $\circ \alpha$ means that α is true (or known) *per se* and $\circ \alpha$ that α is necessary; (VII) is a schema of premises, instead of a single premise.

Adding the following inference rules and axiom to our list of logical principles:

C13. $\alpha \to \beta \vee \varphi, \neg \beta, \neg \varphi \vdash \neg \alpha$

C14. $\neg \Box \neg \alpha \vdash \Diamond \alpha$

C15. $\Diamond \Diamond \alpha \vdash \Diamond \alpha$

we have a full Leibnizian derivation for the conclusion that A and B are compossible — $C(A, B)$ or $\Diamond \exists z(A(z) \wedge B(z))$ — as follows:

17.	$\Box \neg C(A, B) \to \blacksquare \neg C(A, B) \vee \circ \neg C(A, B)$	Pr. (VII)
18.	$\neg \circ \neg C(A, B)$	Pr. (VIII)
19.	$\neg \Box \neg \Diamond \exists z(A(z) \wedge B(z))$	C13 16,17,18
20.	$\Diamond \Diamond \exists z(A(z) \wedge B(z))$	C14 19
21.	$\Diamond \exists z(A(z) \wedge B(z))$	C15 20

The second criticism Oppy offers is a threefold one [27, pp. 25–26]. First, even if one succeeds in showing that all simple, positive, absolute qualities are compatible, it seems there is still a hole in the ontological argument: one has to show that there are indeed simple, positive and absolute qualities. Second, given the nature of simple, positive, absolute qualities, there seems to be an epistemological problem about the possibility of reasonable belief in their existence.[26] Third, even if we grant that there are simple, positive, absolute qualities, the question can be raised whether existence is a simple, positive, absolute quality.

Although Oppy's criticism looks correct, it seems to be at odds with his analysis of the argument, and consequently with my own rendition of it. A good look at the first part of the derivation will suffice for one to see why: as far as premise

---

[26] He writes: "What grounds could one have for thinking that there are simple, positive, absolute qualities? There may only be the appearance of a problem here, since it seems reasonable to allow that reasonable belief need not require grounds. However, this problem does appear to threaten the dialectical value of the demonstration; it certainly seems that one could reasonably believe that there are no simple, positive, absolute qualities." [27, p. 25].

(I) — that every simple quality that is positive and absolute is irresolvable — is concerned, although we have applied it to both *A* and *B*, only the result of applying it to *A* (which got us ¬R(*A*) at step 4) is effectively used in the derivation (step *13); steps 5 to 7 play no role whatsoever in the derivation and could, therefore, be harmlessly erased from it. Hence, as far as Leibniz's argument is concerned, only one of the properties needs effectively to be simple, positive and absolute. Based on this, someone might respond to Oppy's criticism as follows: contrary to what he says, one does not need to show that there *are* simple, positive and absolute qualities; all it is required is that one shows that one, and only one, of the divine properties is simple, positive and absolute, and it does need to be the property of existence.

Although surely sound from the point of view our logical analysis, there seems to be something wrong with this response. That only one of the divine properties needs to me simple, positive and absolute does not match with Leibniz account. The solution of this puzzle is in fact very simple: the analysis on which our formalization was based, Oppy's analysis, is faulty. Suppose that *A* is a simple, positive and absolute quality but *B* is not; while *B* is therefore resolvable, *A* is not. In addition, suppose that *B* is resolved into $\vec{A}$ and *C*, where $\vec{A}$ is the complement of *A*. It can therefore be demonstrated that *A* and *B* are incompatible, even though only *A* is unresolvable. What this shows is that in order to demonstrate that *A* and *B* are incompatible, *A or B* must be resolved, not *A and B*. Therefore, the correct representation of premise (iv) is

iv. In order to demonstrate the incompatibility of *A* and *B*, *A or B* must be resolved.

not "In order to demonstrate the incompatibility of *A* and *B*, *A and B* must be resolved", as Oppy says.

Modifying (IV) accordingly —

(IV) $\forall X \forall Y (\blacksquare \neg C(X,Y) \to R(X) \vee R(Y))$

we get a proper reconstruction of Leibniz' reasoning which is not susceptible to the mentioned response:[27]

1. $\forall Y (P(Y) \to \neg R(Y))$ Pr. (I)

2. $P(A)$ Pr. (II)

---

[27] I am considering here (as Leibniz and Oppy did) only pairs of properties, which is obviously not general enough. For an account which considers not only pairs of properties but a potentially infinite numbers of divine properties, see [28].

3. $P(A) \to \neg R(A)$ C2² 1

4. $\neg R(A)$ MP 2,3

5. $P(B)$ Pr. (III)

6. $P(B) \to \neg R(B)$ C2² 1

7. $\neg R(B)$ MP 5,6

8. $\forall X \forall Y (\blacksquare \neg C(X,Y) \to R(X) \vee R(Y))$ Pr. (IV)

9. $\forall Y (\blacksquare \neg C(A,Y) \to R(A) \vee R(Y))$ C2² 8

10. $\blacksquare \neg C(A,B) \to R(A) \vee R(B)$ C2² 9

*11. $\blacksquare \neg C(A,B)$ Pr. (V)

*12. $R(A) \vee R(B)$ MP 11,10

*13. $\neg(R(A) \vee R(B))$ C16 4,7

*14. $(R(A) \vee R(B)) \wedge \neg(R(A) \vee R(B))$ C12 12,13

15. $\blacksquare \neg C(A,B) \to (R(A) \vee R(B)) \wedge \neg(R(A) \vee R(B))$ C6 11,14

16. $\neg \blacksquare \neg C(A,B)$ C7* 16

where the additional rule of inference C16 is as follows:

C16. $\neg \alpha, \neg \beta \vdash \neg(\alpha \vee \beta)$

Oppy's mistake seems to be a good example of an informal analysis which gives rise to a formal reconstruction that does not meet the similarity criterion and, to the extent that the misrepresentation of premise (iv) gives rise to a fake response to one of Oppy's objections to Leibniz's solution, does not meet with the non-troublemaker criterion.

6 Kant

In his *Critique of Pure Reason* (1787, 2nd edition), Immanuel Kant presents three objections against what he calls "the ontological argument", which is, grossly speaking, the cartesian argument we discussed above. The first two critiques are general ones, addressed not to specific formulations of the ontological argument, but to any *a priori* argument for the existence of God. They follow the same general idea put forward by David Hume in his *Dialogues concerning Natural Religion* (1779):

> I shall begin with observing that there is an evident absurdity in pretending to demonstrate a matter of fact, or to prove it by any arguments a priori. Nothing is demonstrable unless the contrary implies a contradiction. Nothing that is distinctly conceivable implies a contradiction. Whatever we conceive as existent, we can also conceive as non-existent. There is no being, therefore, whose non-existence implies a contradiction. Consequently there is no being whose existence is demonstrable. I propose this argument as entirely decisive, and am willing to rest the whole controversy upon it.[28]

Otherwise said, since no existence claim is contradictory, for its negation is always possible, there cannot be any *a priori* proof for the existence of God or any other matter of fact.

Kant in some sense elaborates on this idea, now making use of the distinction between analytic and synthetic claims, the counterparts of Hume's relations of ideas and matters of fact, respectively. He follows Hume in maintaining that synthetic propositions can never be proved *a priori*; this is a prerogative of analytic propositions. Since existential claims are synthetic, he adds, it follows that no ontological proof of the existence of God is possible:

> If we admit, as every reasonable person must, that all existential propositions are synthetic, how can we profess to maintain that the predicate of existence cannot be rejected without contradiction? This is a feature which is found only in analytic propositions, and is precisely what constitutes their analytic character.[29]

But independently of the analytic-synthetic distinction, and this is Kant's second criticism, it is not difficult to see that negative existentials can never be contradictory. If we deny, say, that God is omnipotent, then we arrive at a contradiction, for we suppose that the property of omnipotence belongs to the very concept of an infinite being. But this is very different from denying that God exists; at this time, the instantiation in reality of the whole concept of God, with all its attributes, is denied, and this implies no contradiction whatsoever:

> To posit a triangle, and yet to reject its three angles, is self-contradictory, but there is no contradiction in rejecting the triangle together with its three angles. The same holds true of the concept of an absolutely necessary being. If its existence is rejected, we reject the thing itself with all

---

[28] Translation by H. Aiken [15, p. 58].
[29] Translation by N. Kemp-Smith [16, A598B626].

its predicates; and no question of contradiction can then arise. There is nothing outside it that would then be contradicted, since the necessity of the thing is not supposed to be derived from anything external; nor is there anything internal that would be contradicted, since in rejecting the thing itself we have at the same time rejected all its internal properties. 'God is omnipotent' is a necessary judgement. The omnipotence cannot be rejected if we posit a Deity, that is, an infinite being; for the two concepts are identical. But if we say, 'There is no God', neither the omnipotence nor any other of its predicates is given; they are one and all rejected together with the subject; and there is therefore not the least contradiction in such a judgement.[30]

Not considering other issues such as the tenability of the analytic-synthetic distinction, these two objections have a very serious flaw: they completely overlook the ontological argument itself, or, in other words, what seems to be the very counter-example to the thesis (present in both objections) that no *a priori* proof of God is possible. Only in his third objection, which is by far the most famous, is that Kant addresses directly the (Cartesian) ontological argument:

'Being' is obviously not a real predicate; that is, it is not a concept of something which could be added to the concept of a thing. It is merely the positing of a thing, or of certain determinations, as existing in themselves. Logically, it is merely the copula of a judgement. The proposition 'God is omnipotent' contains two concepts, each of which has its object — God and omnipotence. The small word 'is' adds no new predicate, but only serves to posit the predicate in its relation to the subject. If, now, we take the subject (God) with all its predicates (among which is omnipotence), and say 'God is' or 'There is a God', we attach no new predicate to the concept of God, but only posit the subject itself with all its predicates, and indeed, posit it as an object that stands in relation to my concept. The content of both must be one and the same; nothing can have been added to the concept, which expresses merely what is possible, by my thinking its object (through the expression 'it is') as given absolutely. [...] By whatever and however many predicates we may think a thing — even if we completely determine it — we do not make the least addition to the thing when we further declare that this thing is. Otherwise, it would not be exactly the same thing that exists, but something more than we thought in the concept; and we could not,

---

[30]Translation by N. Kemp-Smith [16, A595B623].

therefore, say that the exact object of my concept exists. If we think in a thing every feature of reality except one, the missing reality is not added by my saying that this defective thing exists.[31]

This is of course the famous existence-is-not-a-predicate criticism against the ontological argument. It is easy to see how it threatens Descartes' argument: if existence is not an authentic predicate, premise

(ii) Existence is a perfection.

is false, for in order for something to be a perfection it must be a predicate. By implication, it also threatens Sobel's formalization and my amendments of it.

The same however cannot be uncontroversially said about Anselm's formulation. Gareth Matthews, for example, has written as follows:

> He does [Anselm] not speak of adding the concept of existence, or even the concept of existence in reality, to the concept of God, or the concept of something than which nothing greater can be thought. What he does instead is to ask us to compare something existing merely in the understanding with something existing in reality as well. And the second, he says, is greater.[32]

Indeed, the key premise of Anselm's argument — premise (x) — and correlated doctrine (G) neither speak nor presuppose that existence is a property or perfection. Instead, they just make the comparative claim that it is greater to exist in reality than to exist merely in the understanding. So, it is not at all clear that Kant's criticism threatens Anselm's formulation.[33]

Oddly enough, it does threaten Adams' formulation of Anselm's argument. Since Adams represents the concepts of existence in reality and existence in the understanding with the help of logical predicates, *his formulation* naturally assumes that existence is a predicate, which implies the odd fact that while Anselm's formulation is at least defensible against Kant's critique, Adam's formalization of it is not. This is again a clear instance of a troublemaker *explicatum*.

It is important to keep in mind that the choice of representing the two existence concepts as logical predicates is exactly this: a technical choice. Many formalizations of Anselm's argument represent at least one of the concepts with the help of the existential quantifier. And in fact, it is not difficult to conceive an alternative version

---

[31] Translation by N. Kemp-Smith [16, A598B626-A600B628].

[32] [23, p. 90].

[33] For the same reason, Oppy's objection that one has to show that existence is a positive, simple and absolute property does not uncontroversially apply to Anselm's argument.

of Adam's formalization which represents none of the two existence concepts as properties. In order to illustrate this point, let me give a rough and somehow naïve sketch of what this version would look like.

First, we have to build an expanded first-order logic with two existential quantifiers, say, $\exists$ and $\mathfrak{Z}$.[34] While $\exists$ is a broad quantifier ranging over a large domain D, $\mathfrak{Z}$ is a more restricted one ranging over domain $D' \subseteq D$. As far as Anselm's argument is concerned, D contains all objects, be them located in reality or in the understanding (it does not matter here who's understanding); D' contains only objects located in reality. Therefore, while $\exists x P(x)$ means that x exists in reality or in the understanding and has property P, $\mathfrak{Z} x P(x)$ means that x exists in reality and has property P. Given this, we have two abbreviations:

$$\phi(x, m) =_{\text{def}} Q(m, x) \wedge \neg \Diamond \exists y \exists n (G(n, m) \wedge Q(n, y))$$
$$\varepsilon(x) =_{\text{def}} \mathfrak{Z} y (y = x)$$

ϕ is the same as Adam's abbreviation. $\varepsilon(x)$ means that x exists in reality. The premises and conclusion are then represented as follows:

(I) $\exists x \exists m (\phi(x, m))$

(II) $\forall x \forall m (\phi(x, m) \rightarrow \Diamond \varepsilon(x))$

(III) $\forall x \forall m (\phi(x, m) \wedge \neg \varepsilon(x) \rightarrow \neg \Diamond \neg (\varepsilon(x) \rightarrow \exists n (G(n, m) \wedge Q(n, x))))$

(IV) $\mathfrak{Z} x \exists m (\phi(x, m))$

7 Conclusion

The first aim of this paper was to provide a humble and somewhat historical introduction to the ontological argument. This was done by presenting the contributions of Anselm, Gaunilo, Descartes, Leibniz and Kant. Its second aim was to critically examine the enterprise of formally analyzing philosophical arguments and contribute in a small degree to the debate on the role of formalization in philosophy. For this purpose, I scrutinized Adam's formalization of Anselm's ontological argument and his formalization of Gaunilo's criticism against it, Sobel's formalization of Descartes'

---

[34] As far as I am concerned, I could not find any published formalization of such kind of logic. I however assume that building such a kind of logical system would not be a big issue. I base this assumption on two facts: the great diversity of existing multi-modal logics (that is to say, modal logics with more than one pair of modal operators) and the well-known equivalence between first-order logic with one variable and (mono) modal logic.

ontological argument and Oppy's analysis of Leibniz's proof for the possibility of God.

As I anticipated in the Introduction, my focus was mainly on the drawbacks and limitations of these approaches as attempts to analyze existing philosophical arguments. I have not however tried to put formal philosophy on trial. Au contraire, the modest critical analysis I have made in the previous sections and the synthesis I shall lay down here are meant to uncover some (quite trivial, one might say) dangers facing the enterprise of formally analyzing philosophical arguments. Besides, I have punctually called attention to the advantages of formalization (as in the disambiguation of (G) in Anselm's argument) and tried to amend (or show the possibility of amending) the approaches I have analyzed.

In one sense, all my criticisms had to do with Carnap's similarity criterion and my non-troublemaker criterion. For instance, although Adams' informal analysis of Anselm's argument incorporates many important elements of the original formulation, it unjustifiably ignores that Anselm did give an argument for premise (viii), not thus satisfying the similarity criterion. Sobel also slips into that in his informal analysis of Descartes' argument: he neglects Descartes' assumption that there must be a connection between the property of existence and existence itself. As we have seen, this had the consequence of turning Descartes' valid argument into an invalid one, which would classify his formal reconstruction as a troublemaker one. A similar problem occurs with Oppy's analysis of Leibniz's argument. As we have seen, besides being unfaithful to Leibniz's argument, Oppy's misrepresentation of premise (iv) gives rise to a fake response to one of Oppy's own objections to Leibniz's solution.

For the formal representation of arguments, here we also find violations of the similarity and non-troublemaker criteria. For instance, Adams equates "it can be thought that" with "it is possible that", representing premise (x) with the help of a possibility modal operator for which we find no hint in Anselm's original formulation. Furthermore, he does not provide any kind of justification for this theoretical decision of his. As I said, the *explicatum* is expected to depart in several aspects from the *explicadum*; however, for a big departure like this one is expected to provide strong philosophical arguments. Another troubling and in fact troublemaker aspect of Adam's formalization is his use of two predicates to represent the concepts of existence in reality and existence in the understanding. By introducing another element which was absent from Anselm's original formulation, he made the argument vulnerable to a criticism which the original formulation was at least defensible against.

As far as the similarity criterion is concerned, the reconstruction of the inferential steps of the argument seems to be the stage of the formal analysis of an existing argument which involves most foreign elements. The word "reconstruction" applies perhaps even more strongly here than in the other two stages. For instance, despite of correctly representing two structural features of Anselm's argument (the reductio ad absurdum strategy and the movement back and forth from a universal discourse to talk about particulars), Adams' derivation is mostly formed by elements trivially not found in Anselm's text. The same can be said about my formal analysis of the derivations of Descartes' and Leibniz's arguments.

The lessons here seem to be clear. From the informal analysis to the formal representation of the inferential steps, everything is but a reconstruction of the original argument. As such, in some degree or other they will depart from the original formulation. But if the departure is too much — this is the first lesson — it is hard to see how the formal argument at hand might be taken as a formalization of the original argument. As a consequence of this, there will be hardly any hope of effectively contributing to the philosophical debate involving the argument: unless it is uncontroversial that the formalization is a reconstruction of the argument at hand, it is very hard to see how it will shed noteworthy light on the issues involving the argument. The similarity criterion should therefore be seen as an indispensable desideratum of any formal reconstruction.

The second lesson is related to the non-troublemaker criterion. What is the point of formally reconstructing an argument if the reconstruction produces confusing questions and unfruitful issues which are not found in the original formulation of the argument or obscure important aspects of it? Besides being a technical contribution, a formal analysis of an argument must also be a philosophical contribution to the debate over the argument in question; this is my version of Carnap's fruitfulness criterion. Its purposes should be to clarify, to shed some light on the philosophical issues involving the argument, and not to introduce new, and we might say, artificial problems.

The only way to avoid these two complications — this is the third and last lesson — is that the reconstruction be itself a philosophical endeavor. As soon as one starts a formal reconstruction of an existing argument, many theoretical choices will be made. But they should be rationally, philosophically justified. Formal philosophy must still be philosophy. *Ad hoc* and unjustified theoretical decisions should be reduced to the minimum, and when they are unavoidable, attention should be called upon them; a formal reconstruction should be aware of its own limitations. Only if this is done can the formal analysis be a real philosophical contribution. Then perhaps we might have hopes to revitalize formal philosophy and increase its interaction with non-formal philosophy.

References

[1] R. Adams. The Logical Structure of Anselm's Arguments. *The Philosophical Review* 80: 28-54, 1971.

[2] J. Barnes. *The Ontological Argument*. London: Macmillan. Campbell. 1976.

[3] J. Bennett. Annotated edition of Leibniz's "Meditations on Knowledge, Truth, and Ideas". https://www.earlymoderntexts.com/assets/pdfs/leibniz1684.pdf, 2007.

[4] D. Blumenfeld. Leibniz's ontological and cosmological arguments'. In: *The Cambridge Companion to Leibniz*, (ed.) N. Jolley, Cambridge: Cambridge University Press, 1995, p. 353-381.

[5] R. Carnap. *Logical Foundations of Probability* (2nd edition). Chicago: University of Chicago Press, 1962.

[6] M. J. Charlesworth, Trans. *Anselm's Proslogion*. Oxford: Oxford University Press. 1965.

[7] G. Eder and E. Ramharte. Formal Reconstructions of St. Anselm's Ontological Argument. *Synthese* 192: 2791-2825. 2015.

[8] P. Engel. Formal Methods in Philosophy. In: *Proceedings of the 6th European Congress of Analytic Philosophy*, (eds.) T. Czarnecki et al. London: College Publications, 2010.

[9] K. Gödel. *Collected Works, vol. 3*. Oxford: Oxford University Press, 1995.

[10] S. Hansson. Formalization in Philosophy. *The Bulletin of Symbolic Logic* 6: 162-175, 2000.

[11] C. Hartshorne. *The Logic of Perfection*. LaSalle, IL: Open Court, 1962.

[12] C. Hartshorne. *Anselm's Discovery: A Re-Examination of the Ontological Proof for God's Existence*. La Salle, IL: Open Court, 1965.

[13] J. Hick and A. McGill, eds. *The Many-Paced Argument: Recent Studies on the Ontological Argument for the Existence of God*. London: Macmillan, 1967.

[14] L. Horsten and I. Douven. Formal Methods in the Philosophy of Science. *Studia Logica* 89: 151-162, 2008.

[15] D. Hume and H. Aiken, trans. *Dialogues Concerning Natural Religion*. New York: Macmillan, 1948.

[16] I. Kant and N. Kemp-Smith, trans. *The Critique of Pure Reason* (Second Edition). London: Macmillan, 1933.

[17] G. Klima. Saint Anselm's Proof: A Problem of Reference, Intentional Identity and Mutual Understanding. In: Ghita Holmström-Hintikka (ed.), *Medieval Philosophy and Modern Times*, Dordrecht: Kluwer, 2000, p. 69-87.

[18] R. La Croix. *Proslogion II and III: A Third Interpretation of Anselm's Argument*. Leiden: E. J. Brill. 1972.

[19] E. S. Haldane and G. R. T. Ross, trans. *Philosophical Works of Descartes, Vol I*. Cambridge: Cambridge University Press, 1967.

[20] G. W. Leibniz and A. Langley, trans. *New Essays Concerning Human Understanding*. New York: Macmillan, 1896.

[21] D. Lewis. Anselm and Actuality. *Noûs* 4: 175–88. 1970.

[22] N. Malcolm. Anselm's Ontological Arguments. *The Philosophical Review* 69: 41–62. 1960.

[23] G. Matthews. The Ontological Argument. In: *The Blackwell Guide to the Philosophy of Religion*, (ed.) William Mann, Oxford: Blackwell Publishing, 2005, p. 81-102.

[24] R. E. Maydole. The Ontological Argument. In: *The Blackwell Companion to Natural Theology*, (eds.) W. L. Craig and J. P. Moreland, Oxford: Blackwell, 2009, pp. 553–592.

[25] L. Nolan. Descartes' Ontological Argument, *The Stanford Encyclopedia of Philosophy* (Fall 2015 Edition), (ed.) Edward N. Zalta (ed.), https://plato.stanford.edu/archives/fall2015/entries/descartes-ontological/, 2015.

[26] P. Oppenheimer and E. Zalta. On the Logic of the Ontological Argument. In: *Philosophical Perspectives 5: The Philosophy of Religion*, (ed.) James Tomblin, Atascadero: Ridgview Press, 1991.

[27] G. Oppy. *Ontological Arguments and Belief in God*. New York: Cambridge University Press, 1995.

[28] J. Perzanowski. Logiki modalne a filozofia, *Jagiellonian Univ., Cracow*. 1989.

[29] A. Plantinga. *The Nature of Necessity*. Oxford: Oxford University Press, 1974.

[30] W. Rowe. *Philosophy of Religion: An Introduction*. 4th edition. Belmont: Wadsworth, 2006.

[31] J. Sobel. *Logic and Theism*. New York: Cambridge University Press, 2004.

[32] R. Silvestre. On the Logical Formalization of Anselm's Ontological Argument. *Revista Brasileira de Filosofia da Religião* 2: 142-161, 2015

A Mechanically Assisted Examination of Begging the Question in Anselm's Ontological Argument

John Rushby
*Computer Science Laboratory,
SRI International, Menlo Park CA USA*

Abstract

I use mechanized verification to examine several first- and higher-order formalizations of Anselm's Ontological Argument against the charge of begging the question. I propose three different criteria for a premise to beg the question in fully formal proofs and find that one or another applies to all the formalizations examined. My purpose is to demonstrate that mechanized verification provides an effective and reliable technique to perform these analyses; readers may decide whether the forms of question begging so identified affect their interest in the Argument or its various formalizations.

1 Introduction

I assume readers have some familiarity with St. Anselm's 11'th Century Ontological Argument for the existence of God [2]; a simplified translation from the original Latin of Anselm's *Proslogion* is given in Figure 1, with some alternative readings in square parentheses. This version of the argument appears in Chapter II of the Proslogion; another version appears in Chapter III and speaks of the *necessary* existence of God. Many authors have examined the Argument, in both its forms; in recent years, most begin by rendering it in modern logic, employing varying degrees of formality. The Proslogion II argument is traditionally rendered in first-order logic while propositional modal logic is used for that of Proslogion III. More recently, higher-order logic and quantified modal logic have been applied to the argument of Proslogion II. My focus here is the Proslogion II argument, represented completely

---

I am grateful to Richard Campbell of the Australian National University for stimulating discussion on these topics, to my colleagues Sam Owre and N. Shankar for many useful conversations on PVS and logic, and to the anonymous reviewers for very helpful comments.

> We can conceive of [something/that] than which there is no greater
>
> If that thing does not exist in reality, then we can conceive of a greater thing—namely, something [just like it] that does exist in reality
>
> Thus, either the greatest thing exists in reality or it is not the greatest thing
>
> Therefore the greatest thing exists in reality
>
> [That's God]

Figure 1: The Ontological Argument

formally in first- or higher-order logic, and explored with the aid of a mechanized verification system. Elsewhere [26], I use a verification system to examine renditions of the argument in modal logic, and also the argument of Proslogion III.

Verification systems are tools from computer science that are generally used for exploration and verification of software or hardware designs and algorithms; they comprise a specification language, which is essentially a rich (usually higher-order) logic, and a collection of powerful deductive engines (e.g., satisfiability solvers for combinations of theories, model checkers, and automated and interactive theorem provers). I have previously explored renditions of the Argument due to Oppenheimer and Zalta [19] and Eder and Ramharter [12] using the PVS verification system [23, 24], and those provide the basis for the work reported here. Benzmüller and Woltzenlogel-Paleo have likewise explored modal arguments due to Gödel and Scott using the Isabelle and Coq verification systems [5, 6].

Mechanized analysis confirms the conclusions of most earlier commentators: the Argument is valid. Attention therefore focuses on the premises and their interpretation. The premises are *a priori* (i.e., armchair speculation) and thus not suitable for empirical confirmation or refutation: it is up to the individual reader to accept or deny them. We may note, however, that the premises are consistent (i.e., they have a model), and this is among the topics that I previously subjected to mechanized examination [23] (as a byproduct, this examination demonstrates that the Argument does not compel a theological interpretation: in the exhibited model, that "than which there is no greater" is the number zero).

The Argument has been a topic of enduring fascination for nearly a thousand years; this is surely due to its derivation of a bold conclusion from unexceptionable premises, which naturally engenders a sense of disquiet: "The Argument does not, to a modern mind, seem very convincing, but it is easier to feel that it must be fallacious than it is to find out precisely where the fallacy lies" [28, page 472]. Many commentators have sought to identify a fallacy in the Argument or its interpretation (e.g., Kant famously denied it on the basis that "existence is not a predicate"). One direction of attack is to claim that the Argument "begs the question"[1]; that is, it essentially assumes what it sets out to prove [22, 30]. This is the charge that I examine here.

Begging the question has traditionally been discussed in the context of informal or semi-formal argumentation and dialectics [3, 4, 29, 31, 32, 33], where the concern is whether arguments that beg the question should be considered fallacious, or valid but unpersuasive, or may even be persuasive. Here, we examine question begging in the context of fully formal, mechanically checked proofs. My purpose is to provide techniques that can identify potential question begging in a systematic and fairly unequivocal manner. I do not condemn the forms of question begging that are identified; rather, my goal is to highlight them so that readers can make up their own minds and can also use these techniques to find other cases.

The paper is structured as follows. In the next section, I introduce a strict definition of "begging the question" and show that a rendition of the Argument due to Oppenheimer and Zalta [19] is vulnerable to this charge. Oppenheimer and Zalta use a definite description (i.e., they speak of "*that* than which there is no greater") and require an additional assumption to ensure this is well-defined. Eder and Ramharter argue that Anselm did not intend this interpretation (i.e., requires only "*something* than which there is no greater") [12, Section 2.3] and therefore dispense with the additional assumption of Oppenheimer and Zalta. In Section 3, I show that this version of the argument does not beg the question under the strict definition, but that it does so under a plausible weakening. In Section 4, I consider an alternative premise due to Eder and Ramharter and show that this does not beg the question under either of the previous interpretations, but I argue that it is at least as questionable as the premise that it replaces because it so perfectly discharges the main step of the proof that it seems reverse-engineered. I suggest a third interpretation for "begging the question" that matches this case. In Section 5, I consider the higher-order treatment of Eder and Ramharter [12, Section 3.3] and a variant derived from Campbell [7]; these proofs are more complicated than the first-order treatments but I show how the third interpretation for "begging the question" applies to them. I compare these interpretations to existing, mainly informal, accounts of what it means to "beg the question" in Section 6. Finally, in Section 7, I discuss some limitations and possible implications of this work.

---

[1]This phrase is widely misunderstood to mean "to invite the question." Its use in logic derives from medieval translations of Aristotle, where the Latin form *Petitio Principii* is also employed.

2 Begging the Question: Strict Case

"Begging the question" is a form of circular reasoning in which we assume what we wish to prove. It is generally discussed in the context of informal argumentation where the premises and conclusion are expressed in natural language. In such cases, the question-begging premise may state the same idea as the conclusion, but in different terms, or it may contain superfluous or even false information, and there is much literature on how to diagnose and interpret such cases [3, 4, 29, 31, 32, 33]. That is not my focus. I am interested in formal, deductive arguments, and in criteria for begging the question that are themselves formal. Now, deductive proofs do not generate new knowledge—the conclusion is always implicit in the premises—but they can generate surprise or insight; I propose that criteria for question begging should focus on the extent to which either the conclusion or its proof are "so directly" represented in the premises as to vitiate the hope of surprise or insight.

The basic criterion for begging the question is that one of, or a collection of, the premises is equivalent to the conclusion. But if some of the premises are equivalent to the conclusion, what are the other premises for? Certainly we must need all the premises to deduce the conclusion (else we can eliminate some of them); thus we surely need all the other premises before we can establish that some of them are equivalent to the conclusion. Hence, the criteria for begging the question should apply *after* we have accepted the other premises. Thus, if C is our conclusion, Q our "questionable" premise (which may be a conjunction of simpler premises) and P our other premises, then Q begs the question if C is equivalent to Q, assuming P: i.e., $P \vdash C = Q$. Of course, this means we can prove C using Q: $P, Q \vdash C$, and we can also do the reverse: $P, C \vdash Q$.

Figure 2 presents Oppenheimer and Zalta's treatment of the Ontological Argument [19] in PVS. I will not describe this in detail, since it is explained at tutorial level [23] and used [24] elsewhere, but remark that the identifiers and constructs used here are from [24] rather than [23]. Briefly, the specification language of PVS is a strongly typed higher-order logic with predicate subtypes. This example uses only first order but does make essential use of predicate subtypes and the proof obligations that they can incur [27]. The uninterpreted type `beings` is used for those things that are "in the understanding" (i.e., "understandable beings"). Note that a question mark at the end of an identifier is merely a convention to indicate predicates (which in PVS are simply functions with return type `bool`). The predicate `God?` recognizes those beings "than which there is no greater"; the axiom `ExUnd` asserts the existence of at least one such being; `the(God?)` is a definite description that identifies this being. PVS generates a proof obligation (not shown here) to ensure this being is unique (this is required by the predicate subtype used in the definition

```
oandz: THEORY
BEGIN

  beings: TYPE
  x, y: VAR beings

  >: (trichotomous?[beings])

  God?(x): bool = NOT EXISTS y: y > x

  re?(x): bool

  ExUnd: AXIOM EXISTS x: God?(x)

  Greater1: AXIOM FORALL x: (NOT re?(x) => EXISTS y: y > x)

  God_re: THEOREM re?(the(God?))

%--------------- Question Begging Analysis ----------------------

  Greater1_circ: THEOREM God_re => Greater1

END oandz
```

Figure 2: Oppenheimer and Zalta's Treatment, in PVS

of the, which is part of the "Prelude" of standard theories built in to PVS) and ExUnd and the trichotomy of > (also from the Prelude)[2] are used to discharge this obligation. The uninterpreted predicate re? identifies those beings that exist "in reality" and the axiom Greater1 asserts that if a being does not exist in reality, then there is a greater being.

The theorem God_re asserts that the being identified by the definite description the(God?) exists in reality. The PVS proof of this theorem is accomplished by the following commands.

PVS Proof
(typepred "the(God?)") (use "Greater1") (grind)

These commands invoke the type associated with the(God?) (namely that it satisfies the predicate God?), the premise Greater1, and then apply the standard automated proof strategy of PVS, called grind. Almost all the proofs mentioned subsequently are similarly straightforward and we do not reproduce them in detail.

---

[2] Trichotomy is the condition FORALL x, y: x > y OR y > x OR x = y.

As first noted by Garbacz [14], the premise `Greater1` begs the question under the other assumptions of the formalization. We state the key implication as `Greater1_circ` (PVS Version 7 allows formula names to be used in expressions as shorthands for the formulas themselves) and prove it as follows.

```
                                                              PVS proof
(expand "God_re")
(expand "Greater1")
(typepred "the(God?)")
(grind :polarity? t)
(inst 1 "x!1")
(typepred ">")
(grind)
```

The first two steps expand the formula names to the formulas they represent, the `typepred` steps introduce the predicate subtypes associated with their arguments (namely, that `the(God?)` satisfies `God?` and that `>` is trichotomous) and the other steps perform quantifier reasoning and routine deductions.

Given that we have proved `God_re` from `Greater1` and vice-versa, we can easily prove they are equivalent. Thus, in the definition of "begging the question" given earlier, C here is `God_re`, Q is `Greater1` and P is the rest of the formalization (i.e., `ExUnd`, the definition of `God?`, and the predicate subtype `trichotomous?` asserted for `>`).

3 Begging the Question: Weaker Case

Eder and Ramharter [12, Section 2.3] claim that Anselm's Proslogion does not employ a definite description and that a correct reading is "*something* than which there is no greater." A suitable modification to the previous PVS theory is shown in Figure 3; the differences are that `>` is now an unconstrained relation on `beings`, and the conclusion is restated as the theorem `God_re_alt`. As before, this theorem is easily proved from the premises `ExUnd` and `Greater1` and the definition of `God?`. However, `Greater1` no longer strictly begs the question because it cannot be proved from the conclusion `God_re_alt`.

We can observe, however, that this specification of the Argument is very austere and imposes no constraints on the relation `>`; in particular, it could be an entirely empty relation. We demonstrate this in the theory interpretation `eandr1interp`, where all beings exist in reality, and none are `>` than any other (some may think this describes the real world), and beings are interpreted as natural numbers. PVS generates proof obligations (not shown here) to ensure the axioms of the theory `eandr1` are theorems under this interpretation, and these are trivially true.

```
eandr1: THEORY
BEGIN

  beings: TYPE
  x, y: VAR beings

  >(x, y): bool

  God?(x): bool = NOT EXISTS y: y > x

  re?(x): bool

  ExUnd: AXIOM EXISTS x: God?(x)

  Greater1: AXIOM FORALL x: (NOT re?(x) => EXISTS y: y > x)

  God_re_alt: THEOREM EXISTS x: God?(x) AND re?(x)

%--------------- Question Begging Analysis ---------------------

  Greater1_circ_alt: THEOREM trichotomous?(>)
     IMPLIES God_re_alt => Greater1

  Greater1_circ_alt2: THEOREM (FORALL x, y: God?(x) => x>y or x=y)
     IMPLIES God_re_alt => Greater1

END eandr1

eandr1interp: THEORY
BEGIN

IMPORTING eandr1{{
  beings := nat,
  > := LAMBDA (x, y: nat): FALSE,
  re? := LAMBDA (x: nat): TRUE
}} AS model

END eandr1interp
```

Figure 3: Eder and Ramharter's First Order Treatment, in PVS

Such a model seems contrary to the intent of the Argument: surely it is not intended that something than which there is no greater is so because nothing is greater than anything else. So we should surely require some minimal constraint on > to eliminate such vacuous models. A plausible constraint is that > be trichotomous; if we add this condition, as in Greater1_circ_alt, then the premise Greater1 can again be proved from the conclusion God_re_alt. (Note that the string IMPLIES and the symbol => are synonyms in PVS, we alternate them for readability.) A weaker condition is to require only that beings satisfying the God? predicate should stand in the > relation to others; this is stated in Greater1_circ_alt2 and is also sufficient to prove Greater1 from God_re_alt.

In terms of the abstract formulation given at the beginning of Section 2, what we have here is that the conclusion C can be proved using the questionable premise Q: $P, Q \vdash C$, but not *vice versa*. However, if we augment the other premises P by adding some P_2, then we can indeed prove Q: $P, P_2, C \vdash Q$, and also the equivalence of C and Q: $P, P_2 \vdash C = Q$. Thus, Q does not beg the question C under the original premises P but does do so under the augmented premises P, P_2. We will say that Q *weakly begs* the question, where P_2 determines the "degree" of weakness.

In this example, the question begging premise fails our definition of strict begging because it is used in an impoverished theory, and weak begging compensates for that. Another way a premise can escape strict begging is by being stronger than necessary and one way to compensate for that is to strengthen the conclusion by conjoining some S so that $P, (C \wedge S) \vdash Q$ and $P, Q \vdash (C \wedge S)$. However, it may be difficult to satisfy both of these simultaneously and the first is equivalent to weak begging with $P_2 = S$; hence, we prefer the original, more versatile, notion of weak begging.

Observe that one can always construct a P_2 and thereby claim weak begging; the question is whether it is plausible and innocuous in the intended interpretation, and this is a matter for human judgment.

4 Indirectly Begging the Question

Eder and Ramharter consider Greater1 an unsatisfactory premise because it does not express "conceptions presupposed by the author" (i.e., Anselm) [12, Section 3.2] and says nothing about what it means to be *greater* other than the contrived connection to *exists in reality*. They propose an alternative premise Greater2, which is shown in Figure 4. This theory is the same as that of Figure 3, except that Greater2 is substituted for Greater1, and a new premise Ex_re is added.

It is easy to prove the conclusion God_re_alt from the new premises; they also directly entail Greater1 so there is circumstantial evidence that they are question begging. However, it is not possible to prove Greater2 from God_re_alt and the other premises, nor have I found a plausible augmentation to the premises that

```
eandr2: THEORY
BEGIN

  beings: TYPE
  x, y: VAR beings

  >(x, y): bool

  God?(x): bool = NOT EXISTS y: y > x

  re?(x): bool

  ExUnd: AXIOM EXISTS x: God?(x)

  Ex_re: AXIOM EXISTS x: re?(x)

  Greater2: AXIOM FORALL x, y: (re?(x) AND NOT re?(y) => x > y)

  God_re_alt: THEOREM EXISTS x: God?(x) AND re?(x)

END eandr2
```

Figure 4: Eder and Ramharter's Adjusted First Order Treatment, in PVS

enables this. Thus, it seems that **Greater2** does not beg the question under our current definitions, neither strictly nor weakly, so we should investigate whether some alternative method might expose it to this charge.

When constructing a mechanically checked proof of **God_re_alt** using **Greater2** I was struck how neatly the premise exactly fits the requirement of the interactive proof at its penultimate step. To see this, observe the PVS sequent shown below; we arrive at this point following a few straightforward steps in the proof of **God_re_alt**. First, we introduce the premises **ExUnd** and **Ex_re**, expand the definition of **God?**, and perform a couple of routine steps of Skolemization, instantiation, and propositional simplification.

```
                                                              PVS Sequent A
God_re_alt :

[-1]   re?(x!1)
   |-------
{1}    x!1 > x!2
[2]    re?(x!2)
```

PVS represents its current proof state as the leaves of a tree of sequents (here there is just one leaf); each sequent has a collection of numbered formulas above and below the |--- turnstile line; the interpretation is that the conjunction of formulas above the line should entail the disjunction of those below. Bracketed numbers on the left are used to identify the lines, and braces (as opposed to brackets) indicate this line is new or changed since the previous proof step. Terms such as x!1 are Skolem constants. PVS eliminates top level negations by moving their formulas to the other side of the turnstile. Thus the sequent above is equivalent to the following.

```
God_re_alt :                                          Variant Sequent

[-1]   re?(x!1)
[2]    NOT re?(x!2)
  |-------
{1}    x!1 > x!2
```

We can read this as

```
re?(x!1) AND NOT re?(x!2) IMPLIES x!1 > x!2
```

and then observe that Greater2 is its universal generalization.

PVS has capabilities that help mechanize this calculation. If we ask PVS to generalize the Skolem constants in the original sequent, it gives us the formula

```
FORALL (x_1, x_2: beings): re?(x_2) IMPLIES x_2 > x_1 OR re?(x_1)
```

Renaming the variables and rearranging, this is

```
FORALL (x, y: beings): (re?(x) AND NOT re?(y)) IMPLIES x > y
```

which is identical to Greater2. Thus, Greater2 corresponds *precisely* to the formula required to discharge the final step of the proof.

I will say that a premise *indirectly* begs the question if it supplies exactly what is required to discharge a key step in the proof. Unless they are redundant or superfluous, all the premises to a proof will be essential to its success, so it may seem that any premise can be considered to indirectly beg the question. Furthermore, if we do enough deduction, we can often arrange things so that the final premise to be installed exactly matches what is required to finish the proof. My intent is that the criterion for indirect begging applies only when the premise in question perfectly matches what is required to discharge a key (usually final) step of the proof when the preceding steps have been entirely routine. It is up to the individual to decide what constitutes "routine" deduction; I include Skolemization, propositional simplification, definition expansion and rewriting, but draw the line at nonobvious quantifier instantiation. The current example does require quantifier instantiation: a few steps prior to Sequent A above, the proof state is represented by the following sequent.

```
God_re_alt :                                                    PVS Sequent

{-1}  God?(x!1)
{-2}  re?(x!2)
  |-------
[1]   EXISTS x: God?(x) AND re?(x)
```

The candidates for instantiating x are the Skolem constants x!1 or x!2. The correct choice is x!1 and I would allow this selection, or even some experimentation with different choices, within the "obvious" threshold, though others may disagree.

I claim that the sequent constructed by the PVS prover following routine deductions is a good representation of our epistemic state after we have digested the other premises. If the questionable premise then supplies exactly what is required to complete the proof (by generalizing the sequent), then it appears reverse-engineered, and certainly eliminates any hope of surprise or insight. Hence, I consider it to beg the question.

My description of indirect begging is very operational and might seem tied to the particulars of the PVS prover, so we can seek a more abstract definition. After we have installed the other premises, the PVS sequent is a representation of $P \supset C$. The proof engineering that reveals Q indirectly to beg the question shows that Q is what is needed to make this a theorem, so $\vdash Q \supset (P \supset C)$. But more than this, it is *exactly* what is needed, so we could suppose $\vdash Q = (P \supset C)$ and then take this as a definition of indirect begging.[3] Notice that strict begging implies this definition, but not vice-versa. However, a difficulty with this definition is that the direction $\vdash (P \supset C) \supset Q$ is generally stronger than can be proved. The proof engineering approach to indirect begging can be seen as an operational way to interpret and approximate this definition: we use deduction to simplify $P \supset C$ and then ask whether Q is its universal generalization.

In simple cases, the proof engineering approach is straightforward and makes good use of proof automation, but it may be difficult to apply in more complex proofs where a premise is employed as part of a longer chain of deductions. In the following

---

[3]I am grateful to one of the reviewers for this suggestion.

section I show how careful proof structuring can, without undue contrivance, isolate the application of a premise and expose its question begging character.

5 Indirect Begging in More Complex Proofs

In search of a more faithful reconstruction of Anselm's Argument, Eder and Ramharter observe that Anselm attributes properties to beings and that some of these (notably *exists in reality*) contribute to evaluation of the *greater* relation [12, Section 3.3]. They formalize this by hypothesizing some class P of "*greater*-making" properties on beings and then define one being to be greater than another exactly when it has all the properties of the second, and more besides. This treatment is higher order because it involves quantification over properties, not merely individuals. This is seen in the definition of > in the PVS formalization of Eder and Ramharter's higher order treatment shown in Figure 5. Notice that P is a set (which is equivalent to a predicate in higher-order logic) of predicates on beings; in PVS a predicate in parentheses as in F: VAR (P) denotes the corresponding subtype, so that F is a variable ranging over the subsets of P. A more detailed description of this PVS formalization is provided elsewhere [24].

The strategy for proving God_re_ho is first to consider the being x introduced by ExUnd; if this being exists in reality, then we are done. If not, then we consider a new being that has exactly the same properties as x, plus existence in reality—this is attractively close to Anselm's own strategy, which is to suppose that very same being can be (re)considered as existing in reality. In the PVS proof this is accomplished by the proof step

(name "X" "choose! z: FORALL F: F(z) = (F(x!1) OR F=re?)")	PVS Proof Step

which names X to be such a being. Here, x!1 is the Skolem constant corresponding to the x introduced by ExUnd and choose! is a "binder" derived from the PVS choice function choose, which is defined in the PVS Prelude. This X is some being that satisfies all the predicates of x!1, plus re?. Given this X, we can complete the proof, except that PVS generates the subsidiary proof obligation shown below to ensure that the choice function is well-defined (i.e., there is such an X).[4]

EXISTS (x: beings): (FORALL F: F(x) = (F(x!1) OR F = re?))	PVS TCC

---

[4]This is similar to the proof obligation generated for the definite description used in Oppenheimer and Zalta's rendition: there we had to prove that the predicate in **the** is uniquely satisfiable; here we need merely to prove that the predicate in choose! is satisfiable. The properties of the definite description, the choice function, and Hilbert's ε are described and compared in our description of Oppenheimer and Zalta's treatment [23].

```
eandrho: THEORY
BEGIN

  beings: TYPE

  x, y, z: VAR beings

  re?: pred[beings]

  P: set[ pred[beings] ]

  F: VAR (P)

  >(x, y): bool = (FORALL F: F(y) => F(x)) & (EXISTS F: F(x) AND NOT F(y))

  God?(x): bool = NOT EXISTS y: y > x

  ExUnd: AXIOM EXISTS x: God?(x)

  Realization: AXIOM
    FORALL (FF:setof[(P)]): EXISTS x: FORALL F: F(x) = FF(F)

  God_re_ho: THEOREM member(re?, P) => EXISTS x: God?(x) AND re?(x)

END eandrho
```

Figure 5: Eder and Ramharter's Higher Order Treatment, in PVS

This proof obligation requires us to establish that there is a being that satisfies the expression in the **choose!**; it is generated from the predicate subtype specified for the argument to **choose** and is therefore called a PVS Typecheck Correctness Condition, or TCC [27].

Eder and Ramharter provide the axiom **Realization** for this purpose; it states that for any collection of properties, there is a being that exemplifies *exactly* those properties and, when its variable **FF** is instantiated with the term

$$\{G: (P) \mid G(x!1) \text{ OR } G=re?\},$$

it provides exactly the expression above. In other words, **Realization** is a generalization of the formula required to discharge a crucial step in the proof. Thus, I claim that the premise **Realization** indirectly begs the question in this proof. This seems appropriate to me, because **Realization** says we can always "turn on" real existence and, taken together with ExUnd and the definition of >, this amounts to the desired conclusion.

An alternative and more common style of proof in PVS would invoke the premise Realization directly at the point where name and choose! are used in the proof described here. The direct invocation obscures the relationship between the formal proof and Anselm's own strategy, and it also uses Realization as one step in a chain of deductions that masks its question begging character. Thus, use of name and choose! are key to revealing both the strategy of the proof and the question begging character of Realization. Note that the deductions prior to the name command, and those on the subsequent branch to discharge the TCC should be routine if Realization is to be considered indirectly question begging, but those on the other branch may be arbitrarily complex.

Campbell [8], who is completing a new book on the Argument [7], adopts some of Eder and Ramharter's higher order treatment, but rejects Realization on the grounds that it is false. Observe that we could have incompatible properties[5] and Realization would then provide the existence (in the understanding) of a being that exemplifies those incompatible properties, and this is certainly questionable. A better approach might be to weaken Realization to allow merely the addition of re? to the properties of some existing being. This is essentially the approach taken below.

Campbell's formal treatment [8] differs from others considered here in that he includes more of Anselm's presentation of the Argument (e.g., where he speaks of "the Fool"). The treatment shown in Figure 6 is my simplified interpretation of Campbell's approach, scaled back to resemble the other treatments considered. Campbell adopts Eder and Ramharter's higher order treatment, but replaces Realization by (in my interpretation) the axiom Weak_real which essentially states that if x does not exist in reality, then we can consider a being just like it that does. A being "just like it" is defined in terms of a predicate quasi_id introduced by Eder and Ramharter [12, Section 3.3] and is true of two beings if they have the same properties, except possibly those in a given set D. Observe that the PVS specification writes this higher order predicate in Curried form. Here, D is always instantiated by the singleton set jre containing just re?, so we always use quasi_id(jre).

---

[5]Eder and Ramharter are careful to require that all the greater-making properties are "positive" so directly contradictory properties are excluded, but we can have positive properties that are mutually incompatible [15]. Examples are being "perfectly just" and "perfectly merciful": the first entails delivering exactly the "right amount" of punishment, while the latter may deliver less than is deserved.

```
campbell: THEORY
BEGIN

  beings: TYPE

  x, y, z: VAR beings

  re?: pred[beings]

  P: set[ pred[beings] ]

  F: var (P)

  >(x, y): bool = (FORALL F: F(y) => F(x)) & (EXISTS F: F(x) AND NOT F(y))

  God?(x): bool = NOT EXISTS y: y > x

  ExUnd: AXIOM EXISTS x: God?(x)

  quasi_id(D: setof[(P)])(x,y: beings): bool =
      FORALL (F:(P)): NOT D(F) => F(x) = F(y)

  jre: setof[(P)] = singleton(re?)

  Weak_real: AXIOM
      NOT re?(x) => (EXISTS z: quasi_id(jre)(z, x) AND re?(z))

  God_re_ho: THEOREM member(re?, P) => EXISTS x: God?(x) AND re?(x)

END campbell
```

Figure 6: Simplified Version of Campbell's Treatment, in PVS

A couple of routine proof steps bring us to the following sequent.

```
God_re_ho :

{-1}  P(re?)
  |-------
[1]    EXISTS y: y > x!1
[2]    re?(x!1)
```

Our technique for discharging this is to instantiate formula 1 with a being just like `x!1` that does exist in reality, which we name `X`.

```
(name "X" "(choose! z: quasi_id(jre)(z, x!1) AND re?(z))")
```

The main branch of the proof then easily completes and we are left with the obligation to ensure that application of the choice function is well-defined. That is, we need to show

```
EXISTS (z: beings): quasi_id(jre)(z, x!1) AND re?(z)
```

under the condition `NOT re?(x!1)`. This is precisely what the premise `Weak_real` supplies, so we may conclude that this premise indirectly begs the question.

The higher order formalizations considered in this section have slightly longer and more complex proofs than those considered earlier. This means that the indirect question begging character of a particular premise may not be obvious if it occurs in the middle of a chain of proof steps. Use of the `name` and `choose!` constructs accomplishes two things: it highlights the strategy of the proof (namely, it identifies the attributes of the alternative being to consider if the first one does not exist in reality), and it isolates application of the questionable premise to a context where its indirect question begging character is revealed.

6 Comparison with Informal Accounts of Begging the Question

There are several works that examine the Ontological Argument against the charge that it begs the question. Some of these, including the present paper, employ a "logical" interpretation for begging the question, which is to say they associate question begging with the logical form of the argument and not with the meaning attached to its symbols. Others employ a "semantical" interpretation and find circularities in the meanings of the concepts employed by the Argument prior to consideration of its logical form.

Roth [21], for example, observes that Anselm begins by offering a definition of God as that than which nothing greater can be conceived and then claims that greatness already presupposes existence and is therefore question begging. McGrath [18] criticizes Rowe's analysis and presents his own, which finds circularity in the relationship between possible and real existence. (Kant, who named the Argument, declared that existence is not a predicate [16].) Devine [10] (who was writing 15 years earlier than McGrath but is not cited by him) asks whether it is possible to use "God" in a true sentence without assuming His existence and concludes that it is indeed possible and thereby acquits the Argument of this kind of circularity.

All these considerations lie outside the scope considered here. We treat "greater than," "real existence," and any other required terms as uninterpreted constants, and we assume there is no conflict between the parts they play in the formalized Argument and the intuitive interpretations attached to them. We then ask whether the formalized argument begs the question in a logical sense.

Many authors consider logical question begging in semi-formal arguments. Some consider a "dialectical" interpretation associated with the back and forth style of argumentation that dates to Aristotle's original identification of the fallacy (as he thought of it), while others consider an "epistemic" interpretation in the context of standard deductive arguments. Walton [33] outlines a history of analysis of begging the question, focusing on the dialectical interpretation, while Garbacz [13] provides a formal account within this framework. Walton [31] contends that the notion of question begging and the intellectual tools to detect it are similar in both the dialectical and epistemic interpretations, so I will focus on the epistemic case. The intuitive idea is that a premise begs the question epistemically when "the arguer's belief in the premise is dependent on his or her reason to believe the conclusion" [31, page 241].

Several authors propose concrete definitions or methods for detecting epistemic question begging. Walton [31], for example, recommends proof diagrams (as supported in the Araucaria system [20]) as a tool to represent the structure of informal arguments, and hence reveal question begging circularities. He illustrates this with "The Bank Manager Example":

Manager: Can you give me a credit reference?

Smith: My friend Jones will vouch for me.

Manager: How do we know he can be trusted?

Smith: Oh, I assure you he can.

Our interest here is with formal arguments and as soon as one starts to formalize The Bank Manager Example, it becomes clear that the argument is invalid, for it has the following form.

Premise 1: $\forall a, b : \mathtt{trusted}(a) \land \mathtt{vouch\text{-}for}(a,b) \supset \mathtt{trusted}(b)$

Premise 2: $\mathtt{vouch\text{-}for}(\mathtt{Jones}, \mathtt{Smith})$

Premise 3: $\mathtt{vouch\text{-}for}(\mathtt{Smith}, \mathtt{Jones})$

Conclusion: $\mathtt{trusted}(\mathtt{Smith})$

The invalidity here is stark and independent of any ideas about question begging. Walton describes other methods for detecting question begging in informal arguments but most of the examples are revealed as invalid when formalized. While these methods may be of assistance to those committed to notions of informal argument or argumentation, our focus here is on valid formal arguments, so we do not find these specific techniques useful, although we do subscribe to the general "epistemic" model of question begging, and will return to this later.

Barker [4], building on [3, 29], calls a deductive argument *simplistic* if it has a premise that entails the conclusion; he claims that all and only such (valid) arguments are question begging. Our definition for strict begging includes this case, but also others. For example, Barker considers the argument with premises p and $\neg q$ and conclusion p to beg the question, whereas that with premises $p \vee q$ and $\neg q$, and the same conclusion does not, which seems peculiar to say the least. Both of these are question begging by our strict definition.

Now one might try to "mask" the question begging character of an argument that satisfies Barker's definition by adding obfuscating material, so he needs some notion of equivalence to expose such "masked" arguments. However, it cannot be logical equivalence of the premises because the conjunction of premises is identical in the two cases above, yet Barker considers one to be question begging and the other not. Barker proposes that "relevant equivalence" (i.e., the bidirectional implication of relevance logic [11]) of the premises is the appropriate notion. The examples above are not equivalent by this criterion ($\neg q \supset p$ and $\neg q$ illustrate premises that are equivalent to the second example by this criterion) and so the question begging character of the first does not implicate the second, according to Barker.

As noted, all these examples strictly beg the question by my definition and I claim this is as it should be. Recall that a premise strictly begs the question when it is equivalent to the conclusion, given the other premises. Now, the essence of the epistemic interpretation for begging the question is that truth of the premise in question is difficult to know or believe independently of the conclusion, and I assert that this judgment must be made after we have digested the other premises (otherwise, what is their purpose?). Thus, if $\neg q$ is given (digested), then $p \vee q$ and p are logically equivalent and we cannot believe one independently of the other and $p \vee q$ is rightly considered to beg the question in this context. Barker judges $p \vee q$ and p in the absence of any other premise and thereby reaches the wrong conclusion, in my opinion.

My proposal for strict begging differs from those in the literature but is not unrelated to existing proposals such as Barker's. My proposals for weak and indirect begging depart more radically from previous treatments. I consider a premise to be weakly begging when light augmentation to the other premises render it strictly begging. Human judgment must determine whether the augmentation required is innocuous or contrived and this can be guided by epistemic considerations: if the augmentation is required to establish a context in which the questionable premise(s) are plausible (as in our example of Figure 3, where we certainly intend the $>$ relation to be nonempty), then the questionable premise(s) surely beg the question in the informal epistemic sense as well as in our formal weak sense.

Indirect begging arises when the questionable premise supplies (a generalization of) exactly what is required to make a key move in the proof. Provided we have not applied anything beyond routine deduction, I claim that the proof state (conveniently represented as a sequent) represents our epistemic state after digesting the other premises and the desired conclusion. An indirectly begging premise is typically (a generalization of) one that can be reverse engineered from this state, and belief in such a premise cannot be independent of belief in the current proof state; hence such a premise begs the question in the informal epistemic sense as well as in our formal indirect sense.

The informal epistemic criterion underpins our definitions for begging the question in formal deductive arguments. These identify when a premise may be considered to beg the question, but it is not immediate from these definitions why this should be considered a defect. The conclusion to a deductive argument is always implicit or "contained" in the premises but one source of value or satisfaction can be surprise at the revelation that the premises do indeed entail the conclusion. This is surely one reason for the enduring interest in the Ontological Argument: its premises seem innocuous, yet its conclusion is bold. But when a premise is shown to beg the question, this surprise is seen to be illusory: we already assented to the conclusion when we accepted the premise in question.

Most authors who examine question begging in the Ontological Argument implicitly apply an epistemic criterion, and do so in the context of modal representations of the argument (which are briefly mentioned below). Walton, however, does discuss first-order formulations in a paper that is otherwise about modal formulations [30].

Walton begins with a formulation that is identical (modulo notation) to that of Figure 4. He asserts that the premise `Greater2` (his premise 2) is implausibly strong because it "would appear to imply, for example, that a speck of dust is greater than Paul Bunyan."[6] I would suggest that a better indicator of its "implausible strength" is the fact that it indirectly begs the question, as described in Section 4. Walton then proposes that premise `Greater1` of Figure 3 (his premise 2G) may be preferable but worries that our reason for believing `Greater1` must be something like `Greater2`. It is interesting that Walton does not indicate concern that `Greater1` might beg

---

[6]Paul Bunyan is a lumberjack character in American folklore.

the question, whereas our analysis shows that it is weakly begging, and becomes strictly so in the presence of premises that require a modicum of connectivity in the > relation (recall Sections 2 and 3). Thus, I suggest that the formulations and methods of analysis proposed here are more precise, informative, and checkable than Walton's and other informal interpretations for begging the question.

7 Conclusion

Once we go beyond the "simplistic" case, where the conclusion is directly entailed by one of its premises, the idea of begging the question is open to discussion and personal judgment. A variety of positions are contested in the literature on argumentation and were surveyed in Section 6, but I have not seen any discussion of question begging in fully formal deductive settings.

My proposal is that a premise may be considered to beg the question when it is equivalent to the conclusion, given the other premises (strict begging), or a light augmentation of these (weak begging), or when it directly discharges a key step of the proof (indirect begging). The intuition is that such premises are so close to the conclusion or its proof that they cannot be understood or believed independently of it. I have shown that several first- and higher-order formalizations of the Ontological Argument beg the question, illustrating each of the three kinds of question begging. I suspect that all similar formulations of the Argument are vulnerable to the same charge.

Separately (in work performed after this paper was prepared) [26], I have examined several formulations of the argument in quantified modal logic (including that of Rowe [22], who explicitly accuses the Argument of begging the question, and those of Adams [1] and Lewis [17], who also discuss circularity) and found them vulnerable to the same criticism. The analysis there reveals that modal formulations of the Argument admit delicate choices in how the quantification is arranged and this determines identification of the premises accused of question begging.

Begging the question is not a fatal defect and does not affect validity of its argument; identification of a question begging premise can be an interesting observation in its own right, as may be identification of the augmented premises that reveal a weakly begging one. However, I think most would agree that the persuasiveness of an argument is diminished when its premises are shown to beg the question. Furthermore, revelation of question begging undermines any delight or surprise in the conclusion, for the question begging premise is now seen to express the same idea.

Indirect begging is perhaps the most delicate case: it reveals how exquisitely crafted—one is always tempted to say reverse-engineered—is the questionable premise to its rôle in the proof. To my mind, it casts doubt on the extent to which the premise may be considered analytic in the sense that Eder and Ramharter use the term: that is, something that the author "could have held to be true for conceptual (non-empirical) reasons" [12, Section 1.2(7)].

In a related observation, Eder and Ramharter note that in a deductively valid argument the premises always "contain" the conclusion but, for an argument to be satisfying, they should do so in a *non-obvious* way: the conclusion has to be "hidden" in the premises [12, Section 1.2(5)]. One way of looking at the notions of question begging defined here is that they identify cases where the conclusion is insufficiently well hidden. A legitimate criticism is that the methods employed, particularly for weak and indirect begging, may be too powerful, so that intuitively "well hidden" conclusions are exposed by unreasonably intense scrutiny. Richard Campbell expresses this concern [9] and poses an example derived from Proslogion III. Here, we have two premises

1. "Something-than-which-a-greater-cannot-be-thought" (STWNG) so truly exists that it cannot be thought not to exist.

2. Whatever is other than God can be thought not to exist.

The desired conclusion is "God is STWNG" (and hence exists).

Since this has only two premises, if we are given either one plus the conclusion it is always possible to calculate the other, and Campbell is concerned this can be used to justify an accusation of indirect begging. He finds this argument to be a satisfying one since the first premise says nothing about God, and the second says nothing about His greatness, so the conclusion is nicely hidden in the premises.

This is a modal argument (i.e., it involves *necessary* and *possible* existence) and I prefer not to complicate this paper with a description of how modal arguments are embedded in PVS (this is done at length elsewhere [25, 26]), but the salient point is that my methods can indeed be used unjustly to accuse this argument of begging the question.

First, we might claim that STWNG should be unique under the intended interpretation. If we add this as a premise, then Premise 2 can be proved from the other premises and the Conclusion and is therefore weakly begging. Separately, Premise 1 can be reverse-engineered (and thereby claimed as indirectly begging) from the Conclusion and Premise 2, but the derivation involves a quantifier instantiation (of STWNG for the "whatever" variable in Premise 2).

Now, weak begging is "graduated" by the strength of the augmenting premise, and indirect begging by the deductive power employed, so this example nicely illustrates the range of judgments that are possible. Campbell states that to augment the premises is not merely unnecessary but an error, for uniqueness is a consequence, not an assumption, of this argument. Furthermore, the quantifier instantiation required to exhibit indirect begging is not routine (indeed, PVS does not find it automatically) but a creative step. Thus, the accusations of weak and indirect begging should both be rejected in this example. In contrast, the augmentation and deductive power needed to reveal weak and indirect begging in the Proslogion II argument seem reasonable to me and serve correctly to identify the premises concerned as contrived rather than analytic.

It is, of course, for individual readers to form their own opinions and to decide whether the forms of question begging identified here affect their confidence, or their interest, in the various renditions of Anselm's Argument, or in the Argument itself. What I hope all readers find attractive is that these methods provide explicit evidence to support accusations of question begging that can be exhibited, examined, and discussed, and that may be found interesting or enlightening even if the accusations are ultimately rejected.

Observe that detection of the various kinds of question begging requires exploring variations on a specification or proof. This is tedious and error-prone to do by hand, but simple, fast, and reliable using mechanized assistance. I hope the methods and tools illustrated here will encourage others to investigate similar questions concerning this and other formalized arguments: as Leibniz said, "let us calculate."

References

[1] Robert Merrihew Adams. The logical structure of Anselm's arguments. *The Philosophical Review*, 80(1):28–54, 1971.

[2] St. Anselm. Proslogion. Internet Medieval Sourcebook. Fordham University (in English, the original Latin is dated 1077).

[3] John A. Barker. The fallacy of begging the question. *Dialogue*, 15(2):241–255, 1976.

[4] John A. Barker. The nature of question-begging arguments. *Dialogue*, 17(3):490–498, 1978.

[5] Christoph Benzmüller and Bruno Woltzenlogel Paleo. Gödel's God in Isabelle/HOL. *Archive of Formal Proofs*, 2013.

[6] Christoph Benzmüller and Bruno Woltzenlogel Paleo. Interacting with modal logics in the Coq proof assistant. In *Computer Science—Theory and Applications: 10th International Computer Science Symposium in Russia, CSR 2015*, volume 9139 of *Lecture Notes in Computer Science*, pages 398–411, Listvyanka, Russia, July 2015. Springer-Verlag.

[7] Richard J. Campbell. *Rethinking Anselm's Arguments: A Vindication of his Proof of the Existence of God*. Brill, Leiden, The Netherlands, 2018.

[8] Richard J. Campbell. Personal communication, June 2016.

[9] Richard J. Campbell. Personal communication, February 2018.

[10] Philip E. Devine. Does St. Anselm beg the question? *Philosophy*, 50(193):271–281, July 1975.

[11] J. Michael Dunn. Relevance logic and entailment. In *Handbook of philosophical logic*, pages 117–224. Springer, 1986.

[12] Günther Eder and Esther Ramharter. Formal reconstructions of St. Anselm's ontological argument. *Synthese*, 192(9):2795–2825, October 2015.

[13] Paweł Garbacz. Begging the question as a formal fallacy. *Logique & Analyse*, 177–178:81–100, 2002.

[14] Paweł Garbacz. PROVER9's simplifications explained away. *Australasian Journal of Philosophy*, 90(3):585–592, 2012.

[15] Kenneth Einar Himma. Ontological argument. In James Fieser and Bradley Dowden, editors, *Internet Encyclopedia of Philosophy*. April 2005.

[16] Jaakko Hintikka. Kant on existence, predication, and the Ontological Argument. *Dialectica*, 35(1/2):127–146, 1981.

[17] David Lewis. Anselm and actuality. *Noûs*, 4(2):175–188, May 1970.

[18] P. J. McGrath. The refutation of the Ontological Argument. *The Philosophical Quarterly*, 40(159):195–212, April 1990.

[19] Paul E. Oppenheimer and Edward N. Zalta. On the logic of the Ontological Argument. *Philosophical Perspectives*, 5:509–529, 1991. Reprinted in *The Philosopher's Annual: 1991*, Volume XIV (1993): 255–275.

[20] Chris Reed and Glenn Rowe. Araucaria: Software for argument analysis, diagramming and representation. *International Journal on Artificial Intelligence Tools*, 13(4):961–979, 2004.

[21] Michael Roth. A note on Anselm's Ontological Argument. *Mind*, 79:271, April 1970.

[22] William L. Rowe. The Ontological Argument and question-begging. *International Journal for Philosophy of Religion*, 7(4):425–432, 1976.

[23] John Rushby. The Ontological Argument in PVS. In Nikolay Shilov, editor, *Fun With Formal Methods*, St Petersburg, Russia, July 2013. Workshop in association with CAV'13.

[24] John Rushby. Mechanized analysis of a formalization of Anselm's ontological argument by Eder and Ramharter. Technical Note, Computer Science Laboratory, SRI International, Menlo Park, CA, January 2016.

[25] John Rushby. PVS embeddings of propositional and quantified modal logic. Technical Report, Computer Science Laboratory, SRI International, Menlo Park, CA, June 2017.

[26] John Rushby. Mechanized analysis of modal reconstructions of Anselm's Ontological Arguments. Technical Report, Computer Science Laboratory, SRI International, Menlo Park, CA, 2018. In preparation.

[27] John Rushby, Sam Owre, and N. Shankar. Subtypes for specifications: Predicate sub-

typing in PVS. *IEEE Transactions on Software Engineering*, 24(9):709–720, September 1998.

[28] Bertrand Russell. *History of Western Philosophy: Collectors Edition*. Routledge, 2013.

[29] David H. Sanford. The fallacy of begging the question: A reply to Barker. *Dialogue*, 16(3):485–498, 1977.

[30] Douglas N. Walton. The circle in the Ontological Argument. *International Journal for Philosophy of Religion*, 9(4):193–218, 1978.

[31] Douglas N. Walton. Epistemic and dialectical models of begging the question. *Synthese*, 152(2):237–284, September 2006.

[32] Douglas N. Walton. *Begging the Question: Circular Reasoning as a Tactic of Argumentation*. Greenwood Press, New York, NY, 1991.

[33] Douglas N. Walton. Begging the question as a pragmatic fallacy. *Synthese*, 100(1):95–131, 1994.

A Tractarian Resolution to the Ontological Argument

Erik Thomsen
BlenderLogic
ethomsen@blenderlogic.com

The [ontological] argument does not, to a modern mind, seem very convincing, but it is easier to feel convinced that it must be fallacious than to find out precisely where the fallacy lies.
-Bertrand Russell [27]

Abstract

Ontological arguments for the existence of God highlight classical logic's problematic treatment of the existential entailments of true propositions. Is existence implicitly assumed to hold of the logical subject (i.e., argument) of a true proposition? Or must existence be explicitly predicated? To allow for assertions about non-existent objects, modern non-classical approaches from Meinong to Berto, reject the classical approach traceable to Kant and Frege that associates existence with the logical subjects of true propositions. Yet, in overcoming the acknowledged problems with logical subject-based existential entailment (as exemplified by the existential quantifier) these newer predicate-based approaches have only re-opened the door to the problems created by the ontological argument (e.g., from Anselm, Descartes, Leibniz) which were what originally had motivated Kant to delegitimize existential predicates in the first place. Classical logic's approach to the ontological argument appears to be running in circles.

In this paper, I attempt to simultaneously resolve the problems in both the predicate-based and logical subject-based approaches to the ontological argument (and to the characterization principle-based approaches which are closely related to predicate-based ones) by replacing the notion of existential entailment with the notion of 'sequenced evaluation' as the fundamental entailment that applies to both the logical subject and predicate of a proposition. Towards that end, I use a logic consistent with the principles laid out in Wittgenstein's Tractatus. The recasting of ontological arguments in Tractarian terms appears to show a foundational mistake made by all approaches and how it can be resolved.

The author wishes to thank the reviewers for their detailed suggestions and criticisms that served to greatly improve the quality of the paper.

1 Introduction

An ontological argument for the existence of God begins with one or more aspect(s) of our common intuition intended to be so self-evident that they can be taken as analytic or necessary premises. And then attempts, using only logic as reasoning tool, to conclude that there must exist something in our shared reality that exemplifies or corresponds to the concept of God.

Although there are numerous exemplars [18] and even classifications of ontological arguments (including definitional, conceptual, modal, and mereological [20]), they all rely on some notion of existence in both premises and conclusion. For example, Anselm relies on the existence in reality of something that exemplifies a concept as being greater[1], *in some sense* (having added the generic qualifier 'in some sense' because Anselm does not provide an operational definition of what it means to be greater), than the mere concept itself; Descartes [6] relies on necessary existence being a perfection; Leibniz [13] distinguished existence from essence; Kant [10] relies on existence being implicitly associated with the logical subject of an assertion; Frege [9] equates existence with number, and Plantinga [22] treats existence as a modal possibility.

The notion of existence and how it ought to be associated with the premises and/or conclusions that make up the individual propositions belonging to an ontological argument for God is thus one major problem that must be solved to either construct a successful ontological argument or to successfully demonstrate that no such argument is possible. Either outcome may be called a resolution of the ontological argument.

A second major problem that receives nowhere near the same amount of attention, but which nevertheless must be solved, is understanding how the semantics (or definitions) of key terms (e.g., 'God', 'perfections') can impact the validity, soundness (and triviality) of a logical argument. For example, the mereological argument attributed to Lewis [14] by Oppy [20] rests on a very different semantics for the term 'God' than do most other ontological arguments (whose semantics will be discussed below). By equating God as the mereological sum of what exists, God is stripped of its divine attributes and defined simply in terms of non-sentient existence. From the fact that anything exists (e.g., you now reading this paper) it is trivial to conclude that something exists. In this sense, Lewis' semantics trivializes the ontological argument.

---

[1]"And certainly that greater than which cannot be understood cannot exist only in thought, for if it exists only in thought it could also be thought of as existing in reality as well, which is greater." From Chapter 2 'That God really exists' in [29].

The focus therefore, in this paper, will be on two kinds of ontological arguments that (1) exemplify the wrestling within the logic community over how to associate existence with propositions (e.g., through explicit predication, through logical subjects, or some new way), and that (2) illustrate the importance of term semantics to the construction and characterization of logical arguments and the propositions of which they are composed. Specifically, I will discuss (1) Descartes' reformulation in the Meditations of Anselm's original argument where necessary existence is treated as a valid predicate; and (2) a version of the ontological argument not found in the literature outside of Mion in [16] that I include as a means to highlight the impact of existential generalization on any ontological argument and on existential entailments more generally.[2]

Ontological arguments remain of current interest in large part because they highlight areas where there is still disagreement about relevant logical principles. For example, by rejecting the Kantian hypothesis where existence is entailed by the logical subject of an assertion, (which one might argue had been done in an attempt to refute Anselm's and Descartes' ontological argument) and by treating existence as a real predicate, Priest [23] and Berto [4] now seem to be returning to Descartes' implicit treatment of existence as a real perfection (i.e., predicate) for which the problems in so doing were well understood by Kant. This flip flop is but one indication of a foundational problem having to do with the essence, not of God, but of the components of a proposition (i.e., logical subject and predicate) and with the relationship between logic and the world (a point of contention between Wittgenstein and his view of Russell and Frege ([32] 5.4)) whose implications run through the heart of modern logic, both classical and non-classical.

In this paper, I attempt to resolve both kinds of ontological arguments. (admittedly with the less-than-shocking conclusion that they fail, but with new reasons for why they fail). I will try to do this by showing (1) the fundamental mistakes that occur in both Descartes' argument and any argument from existential generalization; and (2) how these mistakes reflect a foundational problem that lies at the heart of both classical logic and the supposed real predicate fixes of Meinong, Priest and Berto that diverge from the classical tradition.

This will be done by using a logic consistent with the principles laid out in Wittgenstein's Tractatus [7]. These principles include a radical re-interpretation of the components of a proposition that defines logical subjects and functions/predicates in terms of sequenced computational processes instead of as references to general objects and properties.

---

[2]Leibniz's shoring up of Anselm's argument by more precisely justifying the ability to conceive of a perfect being is interesting but outside the scope of this paper which focuses on those aspects of the Ontological argument directly impacted by logic's approach to existence.

Towards that end, the rest of the paper is divided in three sections. In §2, I provide an analysis of Descartes' argument and an illustrative argument from existential generalization and include a critical discussion of Priest's recent claim that existential entailment is irrelevant. In §3, I describe relevant aspects of an alternative approach to logic; what might be called Tractarian logic. And in §4, I attempt to resolve the two contrasting ontological arguments by recasting them in Tractarian terms.

2 Problems With The Ontological Arguments

The two major problems with the ontological arguments—existential implications and term semantics are now discussed.

1. Existential implications are the first problem with ontological arguments. At first sight, ontological arguments highlight logic's reliance on some notion of existence either being implicitly entailed by the logical subject of a true proposition or being explicitly predicated. For Descartes, existence (i.e., necessary existence) was a perfection (i.e., a positive predicate).[3] Consider Descartes argument from his Meditations on First Philosophy; Fifth meditation (p 24):

> Whenever it happens that I think of a first and a sovereign Being, and, so to speak, derive the idea of Him from the storehouse of my mind, it is necessary that I should attribute to Him every sort of perfection, although I do not get so far as to enumerate them all, or to apply my mind to each one in particular. And this necessity suffices to make me conclude (after having recognized that existence is a perfection) that this first and sovereign Being really exists.

Descartes' argument can be put in a more explicitly logical form as follows:

Descartes' Argument

> **Premise (#1):** I have an idea of a supreme being that has all perfections;
>
> **Premise (#2):** Necessary existence is a perfection.
>
> **Conclusion (#3):** A supreme being (i.e., God) necessarily exists.

---

[3]Descartes also thought that whatever could be clearly and distinctly perceived to be contained in the idea of something was true of that thing.

For Kant, the fault in Descartes' argument lay in the second premise. Existence could not be allowed to be a perfection (i.e., a predicate) lest Descartes' conclusion prove true. In his motivation to disallow the existential predicates assumed by Descartes, (and Anselm, Leibniz, and Spinoza [31] before), Kant argued in the Critique of Pure Reason (1787) that existence is analytically presumed in the logical subject and so not a real predicate.[4] Kant's view of the existential import of logical subjects was later baked into the fabric of what is called the existential quantifier in classical predicate or first order logic 'FOL'.

However, from the standard interpretation of the existential quantifier, another kind of ontological argument can arise;[5] one based directly on the classical notion of existential generalization 'EG' where f(a) implies $\exists x\, f(x)$.
(Read: f(a) implies that there exists an 'x' such that f is true of x.)

Consider, now, an EG argument.

(2) An EG Argument

Premise (#1): God is perfect.

Conclusion (#2): There exist an X (e.g., 'God') such that X is perfect.

Since the asserted premise is true by definition, the conclusion is not only valid, but also sound. Therefore, God exists. But something is clearly wrong. This EG argument seems like it should be trivially invalid. It shouldn't be that easy to infer an existential claim for the logical subject of an assertion.

So, it is not surprising that the entailment of existence from the logical subjects of true assertions has been challenged. Beginning in the last century, such non-classical thinkers as Meinong, (in [17]) and more recently Lambert [12], Oppenheimer and Zalta [19], Priest [23] and Berto [4], found the classical, or logical subject-based FOL view of existential entailments to be problematic. There are two aspects to this more modern view of the problem.

The first aspect relates to the EG argument presented above where non-classical thinkers did not wish to see existential claims implicitly attributed to the logical subject of an assertion. Rather, they wanted to be able to freely assert or negate the (real) existence of the logical subject of a true assertion. The second aspect is

---

[4]In On Denoting [28, p. 491], Russell also criticized Descartes' argument. But his criticism stating that Descartes' ontological argument fails for "want of a proof of the premise" was unfair because Descartes treated God's perfection as analytic (coming from 'the storehouse of my mind').

[5]Though not an ontological argument in the sense of an argument widely discussed in the literature, the EG argument is a way of exemplifying the problems that arise from associating the quantification of a variable with any kind of existential entailment which is a significant problem as described elsewhere in this essay.

the classically interpreted FOL's seeming inability to support useful reasoning over fictional objects (e.g., Sherlock Holmes) which is something these same individuals quite understandably wanted to be able to do. And in a formally supported way.

Solving both aspects to the problem, it was argued, required existence to be a real predicate and concomitantly, for the quantifier to free itself of any existential implications. What was hitherto called the existential quantifier is now referred to by some as the particular quantifier; sometimes even given a new symbol as with Lambert's Free Logic [12]. While the recent advocacy for existence as a real predicate seems to have solved the second aspect to the problem of existential entailments, namely to allow useful reasoning over logical subjects that admittedly have no reference, it came at a heavy cost; for it reopened the door to the problems in Descartes' original existence-as-a-predicated-perfection ontological argument. Thus, in the sense of whether existence is associated with the predicate or logical subject of a proposition, we are back to the view initially espoused by Descartes that existence is a real perfection or predicate.[6] The logical approach to existential entailments has come full circle.

2. Term semantics are the second problem with ontological arguments. Though it is common to see analyses of Descartes' argument [30], [25] omit the propositional attitude 'I have an idea that x' in order to focus on the existential implications of a supreme being having all perfections (e.g., [30] pps. 32-39), doing so fundamentally changes the semantics of the argument by placing both the

---

[6]Some authors who follow in the wake of Meinong (e.g., [26], [21], [24], [3]) debate the issue of existential entailments using the term characterization instead of 'predications assumed to be true of, or that are part of the identity of, an object'. Priest (2016), for example, asserts on p. xviii (and then in chapter 4 beginning p. 83 under the heading 'characterization principle' or CP) that "an object has those properties that it is characterized as having". On closer examination, however, this characterization principle reduces to the circular and trivial assertion that an object for which certain predicates are assumed true (i.e., the properties the object is characterized as having) may be assumed to be truthfully predicated with those predicates (i.e., the object has those properties it is characterized as having). Priest himself criticizes CP as being too general, but when it comes to discussing non-existent objects, for example on page 13, he uses the more traditional language of predicates and refers to existence as a special predicate. Regardless whether he eventually recasts the notion of predicatable existence in suitably restricted CP terms, it is clear that he treats existence as something that may or may not be predicated of an object. (E.g., Sherlock Holmes is characterized as a detective which is to say that it is assumed that the predicate 'is a detective' is true of the object Sherlock Holmes. But this characterization is independent of whether Sherlock Holmes exists.) So, although Priest and others use, at times, different terminology, they adhere to the notion that existence is not entailed by the logical subject of a true proposition but rather that existence is associated with an object through predication. Their arguments thus fit into the existence-as-predicate group described in this paper and so are a part of the train of logical approaches to existence that have come full circle.

premises and the conclusion of Descartes ontological argument in the world of synthetic propositions. As shown by Sobel, when recast in synthetic terms, Descartes' first premise (recast on page 32 as 'A supremely perfect being has every perfection.') either trivially presupposes the conclusion or yields an invalid argument depending on whether the first premise is taken to mean that there exists a supreme being with all perfections (from which the conclusion trivially follows) or whether it is taken to mean that if there is a supreme being it would have all perfections (in which case the conclusion does not follow).

In addition to showing the futility of trying to produce a non-trivially valid ontological argument with synthetic premises, Sobel's analysis shows how changes in term semantics can impact a logical argument. Descartes original argument is, I believe, more sophisticated than accounted for by Sobel's synthetic rendering. This is because Descartes' first premise (as per his 5th meditation cited above) is that of an a priori idea in the mind. And Descartes' goal was to demonstrate that from an a priori idea (i.e., a premise) in the mind we can nonetheless deduce the existence of God in reality. Descartes' 'trick' was the inclusion of an existential commitment (necessary existence) as a 2nd premise in a seemingly non-controversial way, that by virtue of the first premise is accorded both an a priori and synthetic status.

The semantics of an ontological argument are also impacted by the definitions of the terms that (remain in and) make up the argument. Traditional approaches to the ontological argument implicitly treat the semantics (or meanings) of salient terms as a part of common intuition and so not requiring explicit treatment. Intuition says that 'God has all perfections.' must be analytic. But nowhere in Anselm's original argument or in Descartes' meditations or in Kant's reflection on the arguments is there any treatment of the semantic relationship between crucial terms like *God* and *perfect*. When asserting the premise that "God has all perfections", what is the presumed relation between God and all perfections? Could God have any imperfections? If so, then the assertion that God has all perfections becomes synthetic. And Descartes' argument becomes invalid.

3 Tractarian Logic

Before introducing the relevant aspects of Tractarian logic, it is useful to briefly summarize the background consensus now called First Order Logic or FOL, against which Wittgenstein was arguing. That consensus can be exemplified in two parts.

First, by Frege's characterization of an assertion[7] in terms of f(a) where 'a'

---

[7]Whether in conjunction with an existential quantifier (e.g., "There exists an X such that f(x)"), or a universal quantifier, (e.g., "For all 'x' f(x)"), Hilbert, Peano, Russell, Carnap, Quine

denotes an N-adic logical subject presumed to refer to an ordered set of n objects and 'f' denotes what Frege called a function (now also called a predicate) presumed to refer to a set of objects possessing whatever is the predicate. Common also nowadays is the notion that for a true proposition, what is referred to by the logical subject is a subset of what is referred to by the predicate (e.g., The assertion 'The book is blue' can be interpreted as stating that the book referred to by the logical subject is a subset of the set of blue things). The referential nature of the 'f' and the 'a' has been a central feature of logic from the time of Aristotle to today.

Second, FOL (and even its non-classical offshoots, e.g., [12], [19], [24]) treat individual declarative expressions (whether called wffs, or sentences) as corresponding to individual assertions/propositions (i.e., with a 1-1 relationship between them).

3.1 Tractarian Depiction of the Relationship Between the Logical Subject and Predicate of a Proposition in Terms of Computational Sequence

In contrast, the Wittgenstein of the Tractatus 'TLP' [32] voiced concerns about the referential interpretation of the components of a proposition f and a (3.333). Wittgenstein also distinguished wff/sentences from propositions (5.4733) and suggested a sequential process of evaluation for distinguishing wff that are propositions from those that are not. Legitimate propositions in this scheme maintain their bivalence (4.023).

The process of sequenced evaluation suggested by Wittgenstein is inherently non-commutative (because, as will be shown, different sequences have different truth conditions) and so breaks with the tradition originally espoused by Boole [5]. Given Boole's profound influence on modern logic, and in order to provide a more tangible reference point for understanding the unconventional aspect to Wittgenstein's approach, I succinctly review Boole's major idea below.

From 'The Laws of Thought', chapter 2 on signs and their laws [5], Boole treats multi-part descriptors as set operations. And in the case of multi-term adjectives or verbs associated with an object, he treats them as intersections. Thus, he writes "If x alone stands for white things and y for sheep, let xy stand for white sheep" [page 28]. This is normally understood in an extensional sense, (often with the help of a Venn diagram) as the intersection of the set of white things and the set of sheep things. According to Boole, the intersection is indifferent to order of operation: xy = yx. In chapter IV division of propositions, he treats propositions with the same

---

and Putnam to name but a few, all used a symbolism based on the notion of a predicate 'f' and N-adic logical subject '(a)' to formally denote and reason about propositions.

methods. He distinguishes logical subjects and predicates. *But there is no treatment of the sequential aspects to computation.*

To explain Wittgenstein's alternative view based on sequenced evaluations, let us now turn to the complex fact, illustrated in Figure 1, of a white sheep.

Figure 1: The fact of a white sheep

In (5.5423), Wittgenstein states that two individuals might see different propositions or *logical arrangements* in the same fact/complex.[8] [9] Since the complex fact is represented by two terms: "sheep" and "white", we can form two distinct propositions: "White(sheep)" abbreviated as 'W(s)' and "Sheep(white)" abbreviated as 'S(w)', from the one sentence or wff 'The sheep is white'.[10] Since, as shown in the detailed example below, the truth conditions for these two assertions are different, the issue cannot be brushed off as mere surface grammar. Consider Table 1 (below).

Table 1 specifies (1) two alternate purported propositions 'pp' that can be generated from the one sentence/wff 'The sheep is white' (labeled as purported propositions because we cannot assume they will evaluate); (2) a series of facts beginning with the fact pictured in figure 1 corresponding to row 1 of the table; (3) a collection of related facts in rows 2-4; (4) in the cells that form the intersection of a fact and a

---

[8]The complexity of the fact, what Wittgenstein calls "logical multiplicity" (4.04) governs the number of distinct propositions that can be generated for a single fact/complex.

[9]Wittgenstein's views on the multiplicity of propositions that can be generated from a single complex fact can be traced back to Aristotle's original dialectic ([2, 20_b 22-31], [1, 2_a 4-10, 13_b 10-12]) which begins by looking at an assertion as the answer to a question; not simply as a declarative statement. Aristotle situated logic within the context of an affirming/denying game (the "dialectic", in its original sense), and defined assertions (i.e., propositions) as the primitive units of this game. He further diagnosed a certain compositeness of type as their defining character, distinguishing that which an assertion was asserting from that of which the assertion was being made.

[10]The presence of a name ("Sheep") where a predicate generally occurs, and the presence of a predicate ("white") where a name generally occurs in "Sheep(white)" will be dealt with soon.

	The specific fact (verbally represented)	Purported Proposition pp #1 W(s)	Purported proposition pp #2 S(w)
1	A ***white*** ***sheep***	True proposition	True proposition
2	A blue ***sheep***	False proposition	Logical subject not found; predicate is unevaluable.
3	A ***white*** cow	Logical subject not found; predicate is unevaluable.	False proposition
4	A blue cow	Logical subject not found; predicate is unevaluable.	Logical subject not found; predicate is unevaluable.

Table 1: Truth Conditions of Purported Propositions

pp, the evaluation of the pp when applied to the fact; fact elements that match pp elements (in either pp) are highlighted in ***bold italics***.

Studying Table 1, notice that the facts specified in rows 2-4 diverge from the fact in row 1 that made both pp true propositions. Note also that the two pp differ in terms of the conditions by which their status changes from True to False or from True to unevaluable. Thus in row 2, the fact varies by one element from the original fact: The sheep is blue instead of white. For pp #1, this makes the pp a false proposition. This is because the logical subject, namely "sheep", still matches the fact. But the factual attribute is *blue* not *white* so the pp which asserts that the attribute is *white* is a false proposition. In contrast for pp #2, its logical subject no longer matches an element of the fact. Nothing corresponding to the logical subject 'white' is found. So the predicate "sheep" never gets to evaluate and the pp as a whole is considered unevaluable, because the logical subject fails to match any part of the fact. This is because from pp #2, we cannot find an instance of *white* from where we can evaluate the predicate that identifies what is white. In row 3 the original fact is altered by a different element namely the identity of the object that is white. So here pp #1 becomes unevaluable because its logical subject, "sheep", cannot match any element of the fact; whereas pp #2, which had been unevaluable relative to fact 2, is now simply a false proposition. In row 4, the original fact is completely altered. And neither of the two pp are evaluable.

Since the same facts that make some pp false propositions make others unevaluable (and thus not propositions in the classical bi-valent view) and vice versa, we conclude that the sentence as is (e.g., The sheep is white.), is inherently ambiguous; and the logical (dare I say computable) meaning of a sentence can only be captured by a sequenced process of evaluation that (1) differentiates between pp that do and do not evaluate as propositions, and (2) maintains classical bivalency for those that do evaluate as propositions (e.g., pp #1 is a true proposition for fact 1 and a false proposition for fact 2; while pp #2 is a true proposition for fact 1 and a false proposition for fact 3).

It is important to distinguish false and unevaluable; otherwise one would not be able to distinguish between a fact that negates an assertion (e.g., the fact in row 2 for pp #1) and a fact that has no impact on the truth or falsity of the assertion (the fact in row 4 for both pp #1 and pp #2). Stated alternatively, given the proposition "White(sheep)," it is clear how the fact of a blue sheep would negate the assertion: "No, the sheep isn't white; it's blue." But it is not clear why an independent fact such as 'The cow is blue' should negate the proposition "White(sheep):" "No the sheep isn't white. The cow is blue." If the fact that the cow is blue is allowed to negate "the sheep is white," why not the fact that French is the language of France, or $7 + 2 = 9$? Yet, this is exactly what would happen if false is not distinguished from unevaluable. The door would be opened to allow any non-matching fact to negate an assertion.

One could argue that Wittgenstein is doing no more than suggesting a three-valued logic, e.g., Lukasiewicz [15] and Kleene [11]: a wff is either true or false or unevaluable. But this would ignore Wittgenstein's strong commitment to Bivalence (4.023) and his two-phase sequenced evaluation process. For Wittgenstein, true and false depend on the prior establishment of a pp being a genuine proposition (i.e. on the logical subject matching an aspect of the fact thereby allowing for the predicate to evaluate). It would be hard to overestimate the degree to which this sequenced approach to evaluation is a radical departure from conventional approaches to logic and has significant consequences for reasoning over real world information domains ([8]) and, as will be shown, on resolving the ontological arguments previously described. If an assertion is true, the process of evaluating the predicate must succeed for just that location identified by the logical subject which had to previously succeed in matching some aspect of the fact(s).

3.2 The Tractarian Relationship Between Logic and the World and Resulting Explicit Incorporation of Term Semantics

In (3.33) Wittgenstein writes that "in logical syntax the meaning of a sign should never play a role." Thus, for Wittgenstein, logical syntax (or grammar) is orthogonal to semantics, in the sense that categories of logical syntax (e.g., logical subjects or predicates) can be arbitrarily correlated with meanings (e.g., objects, processes, attributes or relations).

As shown above, the single fact of a white sheep could be the veridical source for two distinct propositions: "White(sheep)" and "Sheep(white)". The two propositions are each comprised of an object 'sheep' and an attribute 'white'. Though it may not be standard to treat an attribute as the logical subject and an object as the predicate, no logical rules are violated by doing so. In fact, each proposition represents an answer to a legitimately distinct question. "White(sheep)" answers the question "What is the color of the sheep?". "Sheep(white)" answers the question "What is it that is white?" The fact that *Sheep* and *white* can each figure in the logical subject or predicate role of a proposition, means that neither logical subjects nor predicates refer in isolation to any specific kind of thing in the world (e.g., logical subjects need not refer to some general object).[11] Only when the logical subjects and predicates have been associated with semantic variables (e.g., specific objects and attributes) do the ensuing propositions refer. Thus, there are no semantic (or ontological or real world) implications to the fact that "sheep" is treated as a logical subject in a proposition. For in another it may be a predicate. The propositional roles of logical subject and predicate are thus orthogonal to whatever semantic types are (extra-logically) postulated as comprising the world.

I believe that Wittgenstein's view of propositions highlights not only the orthogonal relationship between logic and semantics (and the world), but also the treatment of logical variables as computable elements based on their type of semantics. (For example, if the logical subject were a region of Earth, its type might be comprised of a longitude, latitude and altitude.)

To use modern software terminology, Wittgenstein saw logic as a strongly typed system for representing and reasoning with information regardless of origin or meaning ([8]). By separating logic from the world, the Tractarian approach is freed from needless existential implications and is capable of fine tuning the expression of ontological commitments based on the specific ontological realization of the computation (e.g., in your head, in your computer or in the remainder of the real world).

4 Resolving the Ontological Debate Using Tractarian Logic

Now it is time to revisit the ontological arguments described in §2 to see how Tractarian Logic provides a clean resolution to both sides of the existential debate. Let's

---

[11] See sections 3.314–3.317: "And the only thing essential to the stipulation is that it is merely a description of symbols and states nothing about what is signified." (3.317)

begin with what was called an EG argument:

(2) An EG Argument

Premise (#1): God is perfect.

Conclusion (#2): There exists an X (e.g., 'God') such that X is perfect.

For a Tractarian approach, there is no issue with the assumption that the premise "God is perfect" is true. Nor is there any issue with treating that truth as analytic. There are implications for how the terms 'God' and 'perfect' are defined to support those assumptions; but not with the assumptions themselves.

Based on the desired conclusion, (the existence of God in the world), it is reasonable to assume that 'God is perfect' has the form Perfect(God). And it is reasonable to assume that what is perfect is defined as a supertype of God. This is because predicating a supertype of a type (e.g., mammal(dog), liquid(water), color(blue)) is true by definition (i.e., is an analytic truth) whereas predicating a subtype of a type (dog(mammal), water(liquid), blue(color)) or predicating a type of an independent type (blue(book), hot(water)) is contingently true.

The Tractarian treatment of the premise as an analytic truth means that the logical subject of the proposition (i.e., God) has been found and the predicate, 'is perfect', successfully evaluates relative to God. But it doesn't mean that God exists in the world. The existential implications follow from where the proposition is evaluated: in the external world of facts or our internal worlds of definitions. The only way for Perfect(God) to have the status of an analytic truth about the world would be for God to have been found through empirical means and for what is perfect to have been found through empirical means and for what is perfect to be—by definition—a supertype of God. Otherwise, the only way to keep the analyticity of the asserted premise 'God is perfect.' is to restrict its evaluation to the internal world of definitions. In this case, the logical subject 'God' is only committed to being found, when evaluated, amongst the definitions. As a result, the conclusion (which, by definition, is about God in the world) would be invalid. In other words, from the premise "The term 'God' is associated with the term 'perfect'", I can infer that the term 'God' exists amongst my definitions. But no existential claim can be made about the world. In this sense, Tractarian logic does not fall into the ontological trap of existential generalization.

Finally, let us turn our attention to Descartes' more sophisticated predicate-based ontological argument.

Descartes' Argument

Premise (#1): I have an idea of a supreme being that has all perfections;

Premise(#2): Necessary existence is a perfection.

Conclusion (#3): A supreme being (i.e., God) necessarily exists.

Tractarian logic has no issue with Descartes' first premise. Nor, does it need to disallow premise #2 (or existential predications more generally) as did Kant. Rather, from a Tractarian perspective, the big mistake in Descartes' argument is implicitly treating premise #2 in two mutually exclusive ways: one way definitional and one way empirical. The definitional way allows him to combine premise #1 and premise #2. The empirical way allows him to combine premise #2 and the conclusion. But no single interpretation of premise #2 can connect to both premise #1 and the conclusion. Let's look at this more closely.

Premise #1 has the form of a propositional attitude. The outer assertion "I have an idea that x," is what allows the inner assertion "A supreme being that has all perfections" to be treated as a definitional truth. (Remove the outer assertion and Descartes' argument falls prey to Sobel's criticism as described earlier.) Descartes is not asserting that a supreme being exists that has all perfections; only that he has the idea of such a being. One could argue that Descartes' having an idea is a contingent (or empirical) statement. And this would be true if made in the 3rd person (e.g., Descartes' mother said that Descartes had an idea about a supreme being. Maybe he did; maybe he didn't). But Descartes is making the assertion in the first person. And, being aware that one has an idea is arguably as immune from doubt as being aware that one feels pain.

Premise #2 can be evaluated in two different ways. Definitionally, there simply needs to be a term for 'necessary existence', a term for 'perfection', a term for all, and terms for individual perfections. Necessary existence can then be associated with one of the individual perfection terms that comprise the more encompassing term 'all perfections'. Premise #2, that necessary existence is a perfection, would then support an analytic evaluation (presumably by Descartes) which would consist in no more than finding the term 'necessary existence' amongst his definitions and testing whether it is defined as an individual perfection. Of course, this analytic interpretation of premise #2 would make it impossible to conclude that a supreme being necessarily exists in the world. A modified version of Descartes' argument that makes clear the definitional interpretation of premise #2 would look as follows:

Premise (#1): I have an idea of a supreme being that has all perfections;

Premise (#2): By my definition, 'necessary existence' is a perfection.

Conclusion (#3): A supreme being (i.e., God) necessarily exists.

The conclusion in this case would be invalid as neither premise #1 or #2 have in any way asserted anything about supreme beings or God in reality.

Alternatively, premise #2 can be synthetically interpreted. As a synthetic premise one would have to find something in the world (and more than just non-sentient being as that, per the discussion of Lewis above, is trivial) whose existence was necessary. Only when found could one then go back to premise #1 and change its status from definitional (or analytic) to empirical (or synthetic) at which point one could justify the conclusion. A modified version of Descartes' argument that makes clear the synthetic interpretation of premise #2 would look as follows

Premise (#1): I have an idea of a supreme being that has all perfections;

Premise (#2): Something with necessary existence (which is a perfection), exists.

Conclusion (#3): A supreme being (i.e., God) necessarily exists.

This interpretation of premise #2 would yield a valid interpretation for Descartes' argument. But premise #2 is now contingent.[12] It cannot combine with premise #1 until it has been empirically determined to be true. Since we have no guarantee of the truth of premise #2, we cannot combine it with premise #1 to support the conclusion. The argument as a whole would be valid but with unknown soundness.

5 Conclusion

This paper demonstrated that literature on the ontological argument has flip flopped on the pivotal point of classical logic's existential implications. And that the failure to resolve the issue can be traced to a foundational flaw shared by both classical and non-classical logics; namely their referential approach to the components of a proposition which includes the tying of existential claims to what is referred to (or assumed to be referred to) by the proposition's terms (e.g., God, perfections, necessary existence) instead of tying it to how the proposition is evaluated (e.g., in the mind's realm of definitions or in our shared external world). This paper then introduced an approach to logic based on Wittgenstein's Tractatus and showed that with its notion of sequenced evaluations and its separation of computation and reference, it appears able to resolve both the predicate- and logical subject-based approaches to the ontological argument.

---

[12]There are numerous ways one could modify Descartes argument to highlight an empirical interpretation for premise #2. The specific way is not important to the argument.

References

[1] Aristotle. Categories, 350 BCE. Translated by E. M. Edgehill at `http://classics.mit.edu/Aristotle/categories.1.1.html`(Last visited July-3-2018).

[2] Aristotle. On Interpretation, 350 BCE. Translated by E. M. Edgehill at `http://classics.mit.edu/Aristotle/interpretation.html`(Last visited July-3-2018).

[3] Giuliano Bacigalupo. *A Study on Existence*. Cambridge Scholars Publishing, 2017.

[4] Francesco Berto. *Existence as a Real Property: The Ontology of Meinongianism*. Springer Netherlands, 2013.

[5] George Boole. *An Investigation of the Laws of Thought: On Which are Founded the Mathematical Theories of Logic and Probabilities*. Dover Publications, New York, 1958.

[6] René Descartes. *Discourse on Method and the Meditations*. PenguinBooks, Harmondsworth, Eng, 1968. Translated by John Veitch.

[7] Author Removed for Blind Review. . 1990.

[8] Author Removed for Blind Review. . 2003.

[9] Gottlob Frege. *The Foundations of Arithmetic*. Northwestern University Press and Basil Blackwell, Great Britain, 2nd edition, 1968. Translated by J. L. Austin.

[10] Immanuel Kant. *Critique of Pure Reason*. Macmillan, London, 2nd edition, 1933. Translated by Norman Kemp-Smith.

[11] Stephen Cole Kleene. *Two Papers on the Predicate Calculus*. American Mathematical Society, Providence, RI, 1952.

[12] Karel Lambert. *Free Logic: Selected Essays*. The Press Syndicate of the University of Cambridge, United Kingdom, 2018.

[13] Gottfried Wilhelm Freiherr von Leibniz. *New Essays Concerning Human Understanding*. MacMillan, New York, 1896. Translated by Alfred Langley.

[14] David K Lewis. Anselm and Actuality. *Noûs*, 4(2):175–188, 1970.

[15] Jan Lukasiewicz. *Selected Works*. Studies in Logic and the Foundations of Mathmatics. North–Holland Publishing Company, Amsterdam, 1970. Translated by O. Wojtasiewicz.

[16] G. Mion. *On Kant's Ambivalence Toward Existence in His Critique of the Ontological Argument*. Journal of Applied Logic. Forthcoming.

[17] Francesca Modenato. Meinong's Theory of Objects: An Attempt at Overcoming Psychologism. *Grazer Philosophische Studien*, 50:87, 1995.

[18] Yujin Nagasawa. *Existence of God*. Routledge, 2011.

[19] Paul E Oppenheimer and Edward N Zalta. On the Logic of the Ontological Argument. *Philosophical Perspectives*, 5:509–529, 1991.

[20] Graham Robert Oppy. *Ontological Arguments and Belief in God*. Cambridge University Press, Cambridge;New York;, 1995.

[21] Michele Paolini Paoletti. Meinong Strikes Again. *Humana Mente: Journal of Philosophical Studies*, 6(25), 2013.

[22] Alvin Plantinga. *The Nature of Necessity*. Oxford University Press on Demand, 1974.

[23] Graham Priest. The Closing of the Mind: How the Particular Quantifier became Existentially Loaded Behind our Backs. *The Review of Symbolic Logic*, 1(1):42–55, 2008.

[24] Graham Priest. *Towards Non-Being*. Oxford University Press, GB, 2nd edition, 2016.

[25] Graham Priest. *Logic: A Very Short Introduction*, volume 29. Oxford University Press, New York;Oxford;, second edition, 2017.

[26] Richard Routley. *Exploring Meinong's Jungle and Beyond*. Research School of Social Sciences, Australian National University, Canberra, 1980.

[27] Bertrand Russell. *A History of Western Philosophy*. Simon and Schuster, New York, N.Y, 1972.

[28] Bertrand Russell. On Denoting. *Mind, New Series*, 14(56):493–479, 1995.

[29] Anselm Saint Archbishop of Canterbury and Gaunilo. *St. Anselm's Proslogion*. Clarendon Press, Oxford, 1965. Translated and Introduced by Charlesworth, M J.

[30] Jordan Howard Sobel. *Logic and Theism: Arguments For and Against Beliefs in God*. Cambridge University Press, Cambridge, 2004.

[31] Benedictus de Spinoza. *The Ethics*. Dover, New York, 1955. Translated by R. Elwes.

[32] Ludwig Wittgenstein. *Tractatus Logico-Philosophicus*. Routledge Ltd, 1977. Translated by D. F. Pears and Brian McGuinness.

On Kant's Hidden Ambivalence Toward Existence in his Critique of the Ontological Argument

Giovanni Mion
Istanbul Technical University, Turkey
gmion22@gmail.com

Abstract

The paper explores Kant's attitude toward existence in the *Critique of Pure Reason*. It has two main goals: first, it argues that Kant's criticisms of the ontological argument might be vitiated by an ambivalence toward existence, and then it attempts to provide a solution to the ambivalence in question. Finally, since my reading of Kant assumes that for him, *existence* is governed by the rule of existential generalization, I also prove the following biconditional: *existence* is not a real predicate *iff* existential generalization is valid.

Keywords: Kant; Ontological argument; Existential generalization.

0 Introduction

The paper has two goals. The first goal is to draw attention to an interpretative puzzle concerning Kant's attitude toward existence in his critique of the ontological argument that is generally overlooked or underappreciated. The second goal is to suggest a possible solution to the puzzle, and accordingly, to suggest a reading of Kant's doctrine that might be exegetically and theoretically satisfactory.

The paper has three main sections. In the first section, I will show that Kant's attitude toward existence in the *Critique of Pure Reason* (CPR) is highly problematic, and, therefore, that his critique to the ontological argument cannot be uncritically accepted. In the second section, I will try to reconcile the various claims that Kant makes about existence. My solution commits Kant to the validity of the rule of existential generalization. Yet, Kant claims only that *existence* is not a real predicate [4, A599/B627]. Accordingly, in the third section, I will prove the following biconditional: *existence* is not a real predicate *iff* existential generalization is valid.

1 On Kant's ambivalence toward existence

According to Kant, any theoretical argument that aims to establish the existence of God is bound to fail, for we cannot know what we cannot intuit, and God is not something that can be intuited. Accordingly, in CPR, in order to undermine all efforts to overcome what he saw as the limits of knowledge, Kant, among other things, challenges the soundness of the ontological argument for the existence of God. Yet, as I am about to show, his critique of the ontological argument hides an ambivalence toward existence that is rarely appreciated.

In CPR A599/B627, Kant makes two noteworthy claims concerning existence and existential propositions:

(a) *Existence* is not a real predicate.

(b) Existential propositions are synthetic.

With (a), Kant wishes to block the following Cartesian version of the ontological argument:

(c) God has all perfections.
Existence is a perfection.
Therefore, God exists.

(c) is based upon Anselm's intuition about the status of existence as a perfection. According to him, God cannot be truly perfect unless he or she existed. However, for Kant, this argument is unsound, for the second premise is false. Since *existence* is not a real predicate, i.e., a concept of something that could add to the concept of a thing, *a fortiori* it is not a perfection.

On the other hand, with (b) Kant wishes to block any attempt to infer an existential proposition from mere concepts or definitions. If existential propositions are synthetic, then they can only be established by intuition (either pure or empirical). Accordingly, they cannot be derived from analytic truths. This implies, among other things, that the following proposition cannot be contradictory, for it is not analytic.

(d) God does not exist.

As Kant puts it:

I ask you: is the proposition, *This or that thing* (which I have conceded to you as possible, whatever it may be) *exists* - is this proposition, I say, an analytic or a synthetic proposition? [...] If you concede, as in all fairness you must, that every existential proposition is synthetic, then how would you assert that the predicate of existence may not be cancelled without contradiction? - since this privilege pertains only in the analytic propositions, as resting on its very character [4, A598/B626].

(a) and (b) are repeated over and over in scholarly books and articles in connection with Kant's philosophy. Yet, it is not immediately clear that they are actually compatible. In 1962, Jerome Shaffer[1] complained that Kant's definition of real predicate (as something 'which is added to the concept of the subject and enlarges it') is "most unfortunate", for it contradicts his claim that existential propositions are synthetic.[2] Take a paradigmatic example of a synthetic (singular) proposition:

(e) Socrates is wise.

(e) is clearly synthetic, for it cannot be established by conceptual analysis. Moreover, 'wise' in (e) is supposed to be a real predicate, in the sense that it adds/enlarges the (individual) concept 'Socrates'. Consider now the following proposition:

(f) Socrates exists.

According to Kant, (f) is also synthetic, for it cannot be established by conceptual analysis, but only by an empirical investigation. This means that the concept of existence is not contained in the concept 'Socrates'. But if the concept of existence is not already contained in the concept 'Socrates', then it is *added* to it. So, it seems that, after all, *existence* does behave like a genuine predicate. In short, in spite of what Kant claims, if existential propositions are synthetic, then *existence* seems to be a real predicate. Conversely, if *existence* is not a predicate, then existential propositions should be analytic. As Ian Proops puts it:

The problem is that if being is not a real predicate, then, by the definition of 'real predicate', it must be a concept that cannot be 'added' to any other concept. But that

---

[1] "What is a 'real' predicate? Kant defines it as something 'which is added to the concept of the subject and enlarges it'. This is a most unfortunate definition for Kant to use, however, since it leads to contradiction with another important doctrine of his, that existential propositions are always synthetic. Synthetic judgments are those which 'add to the concept of the subject a predicate which has not been in any wise thought in it', and if existential judgments are always synthetic then 'exists' must be a predicate which adds to the concept of the subject, in short, a 'real' predicate as defined above" [9, p. 309].

[2] More recently, Nicholas Everitt wrote: "Although, therefore, Kant's name is particularly associated with the assertion that 'exists' is not a real predicate, we will look in vain to him for a cogent defense of that thesis. Indeed, it seems that Kant's own words commit him to denying the thesis as well as to asserting it" [2, p. 52].

seems to mean that it must be analytically contained in every concept. But, if so, existential judgments would have to be analytic [8, p. 9].

2 A possible solution

As I showed in the previous section, Kant's criticisms of the ontological argument seem to generate an ambivalence toward existence, for if *existence* is not a real predicate, then, in contrast to what Kant claims, existential propositions should turn out to be analytic. In this section, I will sketch a possible solution to the challenge raised by Shaffer.

As I argued in the previous section, for Kant, (e) and (f) are both synthetic. On the other hand, for Kant, the following proposition would be analytic:

(g) If Socrates is wise, then Socrates exists.

Suppose that we established that Socrates is wise. Then we do not need any additional research to claim that Socrates exists, for Socrates cannot be wise unless he exists. So, in spite of the fact that for Kant, 'Socrates is wise' and 'Socrates exists' are both synthetic, (g) is analytic, for no empirical research is needed in order to see that it is true. Moreover, for Kant, also the following proposition would be analytic:

(h) If Socrates is wise, then someone is wise.

Again, suppose that we established that Socrates is wise. Then we do not need any additional research to claim that someone is wise, for Socrates cannot be wise unless 'wisdom' is actually instantiated. So, in spite of the fact that for Kant, 'Socrates is wise' and 'someone is wise' are both synthetic, (h) is analytic, for no empirical research is needed in order to see that it is true.

Noticeably, (h) is an instance of the following logical truth:

(i) If Pa, then $\exists x Px$.

According to (i), if Pa is true, then there exists something that is P. For Kant, (i) would also be analytic, and since no other predicate seems to behave as predicted by (i), he rightly concludes that *existence* is not a *real* predicate, in the sense that even if from a grammatical point of view, *existence* might behave like a predicate, in reality, it works in its own unique way.

If my reading of Kant is correct, then it follows that in the relevant sections of CPR [4, A599/B627], Kant is offering a rudimentary defense of the rule of existential generalization:

(EG) Pa.
 Therefore, $\exists x Px$.

This last claim might seem to be extremely controversial, for Kant only claims that *existence* is not a real predicate. Yet, the idea that Kant's treatment of existence anticipated Frege's and Russell's views on quantified propositions has been endorsed by an impressive range of philosophers: including, Ryle, Carnap, Ayer and Quine.[3] However, none of them bothered to argue for it. (I plan to fill this lacuna in the next section).

My solution has the obvious advantage of dissipating the sense of inconsistency in Kant's treatment of existence raised by Shaffer, for clearly there is no contradiction between the following two claims: 'existential propositions are synthetic' and '*existence* is not a real predicate' if the last claim is interpreted as a defense of EG. However, Kant's implicit endorsement of EG allows for the following argument:

(j) God is perfect.
 Therefore, something is perfect.

(j) is an instance of EG. Moreover, the premise seems to be true. So, it seems that we have just proved the existence of a perfect being.

It is not immediately clear how Kant might respond to (j). Since he would assume its premise to be true by definition, he would have to focus on its validity. In general, if we are confronted with an argument that we wish to invalidate, we have two chief strategies: one, we might argue that the logical rules have not being correctly applied; or two, we might attempt to rephrase the premises in order to show that the conclusion does not really follow from them. As far as I can see, Kant might pursue both strategies.

One, he could try to restrict the validity of EG to synthetic propositions only. In this case, since the premise is analytic, the conclusion would not follow. However, someone might rightly object that this solution would be *ad hoc*, for it seems designed for the sole purpose to block (j). Yet, there seems to be some independent evidence for the claim that, for Kant, existential generalization ought to be restricted in the way envisioned. In a highly exegetical paper, without any specific concern about the puzzle raised in this paper, Alberto Vanzo [11] concludes that for Kant, only

---

[3]For a list of quotations, see [7, pp. 86-87]. More recently, Jonathan Bennett introduced the so-called Kant-Frege view: "According to Kant, every existence-statement says about a concept that it is instantiated, rather than saying about an object that it exists. This is an important precursor of the view of Frege that any legitimate existential statement must be built out of propositional atoms of the form 'There is an F', where F stands for a determining predicate" [1, p. 231].

synthetic judgments have existential import. So if Vanzo is right, Kant might argue that EG does not apply to the argument in question.

Two, Kant might also argue that 'God is perfect' is not a categorical proposition, but an hypothetical one:

(k) If God exists, then he or she is perfect.

Since (k) is a conditional proposition, no categorical (existential) conclusion can be derived from it. As a matter of fact, this is the strategy that Kant seems to pursue in CPR:

All the alleged examples are without exception taken only from **judgments**, but not from **things** and their existence. The unconditioned necessity of judgments, however, is not an absolute necessity of things. For the absolute necessity of the judgment is only a conditioned necessity of the thing, or of the predicate in the judgment. The above [analytic] proposition ['a triangle has three angles'] does not say that three angles are absolutely necessary, but rather that under the [sufficient] condition that a triangle exists (is given), three angles also exist (in it) necessarily [4, A594/B622, my additions].

Here Kant is making two distinct moves. First, he claims that 'a triangle has three angles' is a conditional proposition of the following form:

(l) If a triangle exists, then it has three angles.

Second, he implicitly claims that 'God is perfect' and 'a triangle has three angles' have the same logical form. So, since 'a triangle has three angles' and 'God is perfect' are both conditional propositions, he correctly concludes that no categorical (existential) statement can be derived from them.[4]

3 *Existence* is not a real predicate *iff* EG is valid

In the previous section, I contended that for Kant, EG is valid. As we saw, the view counts numerous illustrious supporters, but it also has some notable detractors, including Hintikka [3] and Kripke [5]. In this paper, I will not deal with their objections.[5] Instead, I will straightforwardly prove the following biconditional: *existence* is not a real predicate *iff* EG is valid.

---

[4]Most logicians, today, would reject (l) as a paraphrases of 'a triangle has three angles'. Yet, they would also claim that the proposition in question is a conditional proposition: for all x, if x is triangle, then x has three angles. Needless to say, no categorical (existential) proposition can be inferred from it.

[5]For a critical assessment of Hintikka's take on the so-called "Kant-Frege view", see [6].

Let us start with a proof of the following conditional claim: if *existence* is not a real predicate, then EG is valid; and following James Van Cleve, let us define 'real predicate' (or 'determining predicate') and 'enlarge' as follows:

(i)   P is a *real predicate iff* P enlarges at least one concept C;

(ii) P *enlarges* C *iff* $\Diamond \exists x(Cx \ \& \sim Px)$.[6]

Take the concept *triangle* and the predicate 'rectangular'. Since it is possible for something to be triangular but not rectangular, 'rectangular' is a real predicate that can enlarge the concept of triangle, and, accordingly, narrow down its extension.

By substitution, from (i) and (ii) we can derive:

(iii) P is a real predicate *iff* for at least one concept C, $\Diamond \exists x(Cx \ \& \sim Px)$.

By contraposition, from (iii), we can derive:

(iv) P is not a real predicate *iff* for any concept C, $\sim \Diamond \exists x(Cx \ \& \sim Px)$.

Since '$\sim \Diamond \exists x(Cx \ \& \sim Px)$' is logically equivalent to '$\Box \forall x(Cx \to Px)$', from (iv) we can derive:

(v)   P is not a real predicate *iff* for any concept C, $\Box \forall x(Cx \to Px)$.

So if 'existence' ('E') is a not a real predicate, then it follows that:

(vi) For any concept C, $\Box \forall x(Cx \to Ex)$.

(vi) says that for any concept 'C', necessarily if 'Ca' is true, then *a* exists. If *a* exists, then 'C' is instantiated. Accordingly:

(vii) If Ca is true, then C is instantiated.

(vii) amounts to the rule of existential generalization (EG):

(viii) From Ca we can always infer $\exists x Cx$.

So if *existence* is not a real predicate, then EG is valid.

Now, suppose that EG is valid. In addition, suppose that *existence* were a real predicate. If *existence* were a real predicate, then it would be possible for 'Ca' to be true and 'Ea' to be false (from iii). This implies that it would be possible for 'Ca' to be true, while C is not instantiated. However, if EG is valid, it would not be possible to have 'Ca' and '$\sim \exists x Cx$'. Accordingly, if EG is valid, then *existence* is not a real predicate.

Q.E.D.

---

[6] See [10, p. 188 and p. 190].

4 Conclusion

This paper had two main goals. Its first goal was to draw attention to an interpretative puzzle concerning Kant's attitude toward existence, and its second goal was to suggest a reading of Kant's doctrine that was exegetically and theoretically satisfactory. More specifically, the challenge was to reconcile the following two claims: existential propositions are synthetic, and *existence* is not a real predicate. If the latter claim is interpreted as a defense of EG, then it becomes clear that there is no contradiction between the following two claims: '$\exists x P x$' is synthetic, and '$Pa \to \exists x P x$' is analytic.[7]

5 References

[1] J. Bennett, Kant's Dialectics, Cambridge University Press, New York, 1974.

[2] N. Everitt, The Non-Existence of God, Routledge, London, 2004.

[3] J. Hintikka, Kant on existence, predication, and the ontological argument, Dialectica 35 (1981) 127-146.

[4] I. Kant, Critique of Pure Reason, P. Guyer and A. W. Wood (Eds.), Cambridge University Press, Cambridge, 1998.

[5] S. Kripke, Reference and Existence, Oxford University Press, New York, 2013.

[6] G. Mion, Hintikka on the "Kant-Frege view": a critical assessment, Logica Universalis (forthcoming).

[7] A. Orenstein, Existence and the Particular Quantifier, Temple University Press, Philadelphia, 1978.

[8] I. Proops, Kant and the ontological argument, Noûs 49 (2015) 1-27.

[9] J. Shaffer, Existence, predication and the ontological argument, Mind 71 (1962) 307-325.

[10] J. Van Cleve, Problems from Kant, Oxford University Press, New York, 1999.

[11] A. Vanzo, Kant on existential import, Kant Review 19 (2014) 207-232.

---

[7] In addition to the referees, I wish to thank John N. Martin, Luis Estrada-González and Erik Thomsen.

The Totality of Predicates and the Possibility of the Most Real Being

Srećko Kovač
Institute of Philosophy, Zagreb,
a public research institute of the Republic of Croatia

Abstract

We claim that Kant's doctrine of the "transcendental ideal of pure reason" contains, in an anticipatory sense, a second-order theory of reality (as a second-order property) and of the highest being. Such a theory, as reconstructed in this paper, is a transformation of Kant's metatheoretical regulative and heuristic presuppositions of empirical theories into a hypothetical ontotheology. We show that this metaphysical theory, in distinction to Descartes' and Leibniz's ontotheology, in many aspects resembles Gödel's theoretical conception of the possibility of a supreme being. The proposed second-order modal formalization of the Kantian doctrine of the "transcendental ideal" of the supreme being includes some specific features of Kantian general logic like "categorical", "hypothetical", and exclusive disjunctive propositions as basic forms.

Keywords: totality of predicates, possibility, most real being, ontological proof, Descartes, Leibniz, Kant, Gödel

1 Introduction

Metalogical and metaphysical concepts are deeply interconnected. This can be shown by the example of the concept of "*all* first-order predicates of things", which is a metalogical presupposition of first-order (general or applied) logic, and of the concept of the "most real being" as the central concept of ontotheology (i.e., ontologically founded philosophical theology). We restrict our considerations to the case of one-place predicates (i.e., concepts in a narrow sense), since such predicates suffice for

---

I am grateful to the reviewers for many valuable comments and suggestions. I am also grateful to the audiences of the *2nd World Congress on Logic and Religion* (Warsaw, June 18–22, 2017) and of the conference *Božje postojanje i Božji atributi* [The existence and attributes of God] (Zagreb, June 23–24, 2016), where earlier drafts of this paper were presented, for helpful discussion and comments.

establishing the concept of the most real being. If we look back into the history of the problem, it seems that there could be a way a metatheoretical totality of predicates of things can be reduced to a single thing, at first, assumed just in the idea, to which all the positive ("real") predicates would belong. We can find such a proposal in Kant's considerations about the "ideal of pure reason". Kant gave these considerations only a provisory and problematic meaning (as if pertaining to "things in themselves", not just to empirical objects), but with a methodological, "regulative" and "heuristic", character for the sake of the "systematic unity" of empirical knowledge (for example, [16, B 713–714]). On the other side, attempts have been made up to the modern times (in so-called ontological arguments) to prove the existence of a being comprising all positive predicates. This opens up the question in which sense, if at all, such an all-comprising ideal, at first just a metatheoretically assumed being, could eventually be an object of a special, metaphysical theory.

We claim that Kant's (critical) account of ontotheology contains the essential contours of an axiomatic description of the concept of a "real" ("positive") predicate, in a significantly similar way as was later formally worked out by Gödel in his axiomatized ontotheology [12]. Kant's account of real predicates ("reality", "thinghood"[1]) does not offer an explicit definition of a real predicate but only a *description* of how real predicates behave in relationship to one another and to other predicates. In distinction, the concept of highest reality (included in each ontotheology) seems to allow Kant a sort of explicit definition of highest being as determined by all real predicates. However, according to Kant, the highest reality neither contains all realities *under* itself – it is not a general concept [16, B 605], nor *in* itself – it is not a whole or a set of realities (second-order domain). To characterize it by means of the totality (*Inbegriff*) of all real predicates (see [16, B 601–602]) is just Kant's first, general approximation to its concept. As we will see, the highest reality should be conceived, more precisely, as the *ground* that contains all realities as its *consequents* [16, B 607].

Regarding the presuppositional character of the highest reality,[2] and the consistency of its fitting into a theoretical, logically built system, the question arises about the *possibility* (logical and real) of the most real being. In particular, whether the most real being, instead of having only a metatheoretical and "regulative" character, can be included into some theory (and which sort of theory) as its proper object, ultimately depends on the answer to the question about the *real* possibility (that is, "objective reality", see, e.g., [16, B 268]) of the most real being. Kant's answer is that the most real being, for a theoretical science, is not possible, not even as a

---

[1]The term "reality" is used by Kant sometimes collectively, for all real predicates taken together, and sometimes distributively, for each real predicate in particular.

[2]Cf. Kant's distinction between relative and absolute supposition in [16, B 704, cf. B 713].

hypothesis in a strict sense [16, B 608, 698] (see footnote 13 below), since it cannot be encountered in sensible experience, although he mentions that it is not "absolutely impossible" *outside* the field of sensible experience [16, B 629, 669]. Moreover, Kant methodologically emphasizes the need of the idea of the most real being as a non-cognizable object presupposed in a relative sense (see "object in the idea" [16, B 698, 725]): in order for us to have a guidance in directing our knowledge towards the ideal goal of a complete cognition of an empirically given object with respect to *all* possible predicates [16, B 608–610, 713–714, cf. B 704–710].

In this paper, we first briefly discuss Kant's result that Cartesian ontological proof suffers from the non-distinguishing of real and logical predicates, which leads to an inconsistency in Cartesian ontotheology, and from taking the possibility of the highest being for granted. We continue with a discussion of Kant's conclusion that Leibniz's proof of the possibility of the highest being does not succeed in demonstrating its real possibility, and show that this is connected with Leibniz's concept of "perfection" as simple. We then propose an explication and formalization of Kant's concepts of a real predicate and of the most real being, indicating how they, in some essential aspects, resemble the concepts of "positive property" and "God", respectively, of Gödel's ontotheology.

Although in Kant's theoretical philosophy, as mentioned above, the highest being is present only in the sense of a methodological presupposition, we claim that Kant's account of "reality", on a "dialectical" extension of the domain of things to "things in themselves" (that is, beyond Kant's "critical" reduction of things merely to sensible objects), suffices to prove the real possibility of the highest being within an appropriate ontotheological theory.[3]

---

[3] See [16, B 602 and B 604] ("Through this possession of all reality ... there is also represented the concept of a *thing in itself* which is thoroughly determined", B 604). This extension corresponds to Gödelian non-sensible intuition (mathematical, conceptual) as a possible means of a *mediate* "knowledge of objects" (cf. [10, p. 268]). For Gödel, Kantian ("concrete") intuition is too weak since it is not sufficient, for example, to ground some elementary or fundamental arithmetical beliefs: "Kant's considerations of pure intuition fail to produce a well-grounded belief in the consistency of arithmetic. This is a ground for rejecting Kant. Our intuition tells us the truth of not only 7 plus 5 being 12 but also [that] there are infinitely many prime numbers and [that] arithmetic is consistent" [35, 7.1.12 p. 217]. On the ground of his logical-mathematical results, Gödel opts for the "idealization" of our intuition up to the intuition of concepts: "Understanding a primitive concept is by abstract intuition" [35, 7.1.13 p. 217]. See also [10, pp. 268–269]. We note that Kant allows for *noumena* ("beings of understanding") in the "problematic" sense of things that can be merely thought of in a non-contradictory way, as well as in the negative sense of things that are not objects of our sensible intuition. However, he does not accept *noumena* in a positive sense of objects of some "non-sensible" ("intellectual") intuition since, according to him, we can make no determinate "concept of a possible intuition" beyond our sensibility [16, B 307–309, 310]. Nevertheless, concepts are for Kant closely interconnected with intuition: it is only by a "schema" in intuition that concepts

2 Descartes and Leibniz

2.1 Descartes

One of the problems Kant addresses is whether there could be any contradiction, in any case, in denying the existence of an object (be it the most real being). As is well known, Kant denies this. He comes to this question while examining the "ontological (Cartesian) proof". In a slightly formalized reconstruction, Descartes' argument reads as follows: if $\forall X(\mathscr{P}erf(X) \to X(x))$ ("to have all perfections", $\mathscr{P}erf$ for 'perfect') is the essence of the "most perfect being" x, and if $\mathscr{P}erf(E)$ ("existence is a perfection", E for 'exists') is an axiom, then what follows is

$$\Box(\forall X(\mathscr{P}erf(X) \to X(x)) \to E(x)) \tag{1}$$

("It is necessary that x that has all perfections, exists"), but not $\Box E(x)$ ("It is necessary that x exists"). At most (from (1)), if the possession of all perfections necessarily characterize x (as its essence), then x necessarily exists: $\Box \forall X(\mathscr{P}erf(X) \to X(x)) \to \Box E(x)$.[4] The subject of (1) (the referent of the antecedent "to be the most perfect being", i.e., "to have all perfections") is meant conditionally, and only under this condition (that an object is a most perfect being) is the existence of this object stated.[5] Moreover, as Kant remarks, it is a contradiction to require that the subject ("most real being") that is thought, at first, merely as possible should contain in its concept (subject term) the concept of existence [16, B 625]. A concept of a thing expresses just the possibility, and thus cannot be extended by the existence predicate in a content-related way [16, B 626–628]. As Kant argues, there is no sense in conceiving an existential proposition other than as a synthetic one, that is, as exceeding the given concept ("subject term" of the proposition) by stating that its object is given. Accordingly, it is logically possible to negate the existence of any subject.[6]

We note that singular propositions (like 'The most perfect being exists') are in Kant's logic universal [16, B 96], i.e., of the form "all S are P" (SaP) or "No S

---

can be applied to sensibly given objects (for example, the schema of magnitude is number, the schema of substance is "the persistence of the real in time" [16, B 182, 183]), whereas a most abstract concept, an idea, is an "analogue of a schema of sensibility" leading not to knowledge of an object but to some "principle of the systematic unity" of our knowledge [16, B 692–693, 698–699].

[4][5]. See [16, B 624–625] and the formalization by Sobel in [31].

[5]This is similar to the example of a triangle in the proposition "A triangle has three angles" [16, B 621–622].

[6]As Kant explains, even if we allowed the subject concept to have in its content the mark of existence (cf. "... under the condition that I posit this thing as given (existing)" [16, B 622]), the existential proposition about this subject would be a mere tautology either just by referring to the subject concept itself as an existing thought and thing or by affirming the existence on the ground

are P" (SeP), and that the truth of a universal proposition does not claim the existence of the things referred to by its subject term, but takes them only as a condition under which the predicate holds.[7] However, the logical impossibility of the subject makes categorical (subject-predicate) propositions, for Kant, false (like both the proposition 'A square circle is round' and 'A square circle is not round', see [17, p. 341, §52b], cf. [16, B 820–821]).

According to Kant's diagnosis, what is responsible for Descartes' mistaken argument is the mixing-up of "logical" and "real" predicates. Descartes' argument treats the existence predicate, which is just "logical" (not pertaining to the content of a concept), as a content-extending, "real", predicate, and is thus mistakenly included in the content of the subject concept ("most perfect being"). Moreover, the real possibility of a "most perfect" being (the possibility of such an object in experience), as well as its logical possibility (non-contradictoriness of its concept), which are the conditions of the statement of the existence of a most perfect being, are in this argument taken for granted.

2.2 Leibniz

We now address the problem of the logical and real *possibility* of the "most perfect being". Kant criticizes Leibniz's proof of the possibility of the most perfect being for not showing more than its logical possibility. We give a reconstruction of Leibniz's ontological proof (metatheoretical, cf. [28]) as it is presented in his *Quod ens perfectissimum existit* (1676, [26]), focusing only on the part of the proof in which it is shown that it is possible that a most perfect being exists, that is, $\Diamond \exists x \, Perfectissimum(x)$. As for Descartes, a most perfect being is conceived as a "subject of all perfections":

Definition 1. $Perfectissimum(x) =_{def} \forall X (\mathscr{P}erf(X) \to X(x))$.

Leibniz conceives "perfection" ($\mathscr{P}erf$) as simple, positive, absolute (without limits), and thus as "purely positive" (*pure positiva*, cannot be understood by means

---

of the existence predicate as previously merely presupposed to be contained in the subject concept [16, B 625—626].

"...I cannot form the least concept of a thing that, if all its predicates were cancelled, would leave behind a contradiction" [16, B 623].

[7]For example, "God is omnipotent" is a necessary judgment: "omnipotence cannot be cancelled *if* you posit a divinity" (our emphasis). Cf. also "mere positings (realities)" of an analytic judgment (see footnote 10). The judgment "God is not" cancels the subject (its existence, *Dasein*, not its logical possibility) together with its predicates and therefore cannot contradict the subject, since the subject is not presupposed to exist [16, B 623] (on the principle of contradiction, see [16, B 190]).

of negations). Therefore, "perfection" is for him non-analysable and non-definable. Now, he defines the "compatibility" (Comp) of properties as the possibility ($\Diamond$) that some x has these properties. Let $\mathscr{T}$ denote any kind of properties.[8] Thus, the definition of compatibility can be given as follows:

Definition 2. $\text{Comp}(\mathscr{T}) =_{def} \Diamond \exists x \forall X (\mathscr{T}(X) \to X(x))$.

Leibniz gives the example of the compatibility of a pair of properties. The concept of compatibility can be simplified accordingly:

Definition 3. $\text{Comp}(X, Y) =_{def} \Diamond \exists x (X(x) \land Y(x))$.

In a modern form, the argument is as follows:

Assumption (A): $\mathscr{P}erf(X) \land \mathscr{P}erf(Y)$,
- $\neg \text{Comp}(X, Y)$ is not provable (since X and Y are non-analysable; from assumption A),
- $\neg \text{Comp}(X, Y)$ is not true in itself (*non est per se vera*),
- if $\Box \phi$ holds, then ϕ is provable from *other* propositions or is true *in itself* (general proposition),

thus, the following propositions hold:
- $\neg \Box \neg \text{Comp}(X, Y)$,
- $\Diamond \text{Comp}(X, Y)$ [interdefinability of modal operators],
- $\Diamond \Diamond \exists x (X(x) \land Y(x))$ (according to Definition 3),
- $\Diamond \exists x (X(x) \land Y(x))$ [modal theorem $4\Diamond$],
- $\text{Comp}(X, Y)$ (according to Definition 3).

For Leibniz, this way of reasoning is applicable to "more" and "any other such qualities" and leads to the conclusion that *all* perfections are compatible. We note that Leibniz's example can be easily generalized up to the case of any finite set of perfections, that is, to the case of $\text{Comp}(X_1, \ldots, X_n)$. For the case of infinitely many perfections, second-order quantification can be used, assuming (A'): $\forall X (\mathscr{U}(X) \to \mathscr{P}erf(X))$ (instead of A), and obtaining $\neg \text{Comp}(\mathscr{U})$ as not provable (since no X with $\mathscr{U}(X)$ is analysable). By $\neg \text{Comp}(\mathscr{U})$ as not true in itself, $\Diamond \exists x \forall X (\mathscr{U}(X) \to X(x))$ follows, which by Definition 2 gives $\text{Comp}(\mathscr{U})$ and, thus, $\text{Comp}(\mathscr{P}erf)$ as a special case (cf. [28]). Of course, the infinity case could also be justified from $\text{Comp}(X_1, \ldots, X_n)$ if the compactness of compatibility is assumed.

What is crucial in Leibniz's proof is the following reasoning:

---

[8] See Perzanowski's reconstruction with "any non-empty family of perfections" in [28, p. 627].

$\neg\phi$ is neither provable from other propositions nor true in itself $\Longrightarrow$ $\neg\Box\neg\phi \Longrightarrow \Diamond\phi$ ($\Longleftrightarrow$ also holds).

For Kant, the possibility of a most perfect being as shown by Leibniz[9] is merely analytic, that is, logical (strictly, we should exclude "existence" from "perfections", since already this is, as mentioned above, a contradiction for him).[10] However, the real possibility of a thing itself, as Kant shows, is not just analytic (in the sense of the provability or truth from concepts), but has a "synthetic" character, that is, it should be ascertained by experience,[11] which is not accomplished by Leibniz's proof.[12]

Thus, what is at stake after these Kantian considerations is to show the *real* possibility of a most perfect being. This is for Kant impossible if what we understand as real possibility is accordance with the conditions of our sensible intuition, since then, in addition to non-contradictoriness, the most perfect being should be possible as an "object of our senses" [16, B 610, 624]. However, if by "most real being" we understand an "object in the idea", then the presupposition of this object seems to be methodologically acceptable and needed.[13] Such a presupposition, as Kant emphasizes, has a regulative and heuristic role in obtaining a "systematic unity" of our knowledge. It is not a constituent part of knowledge, but, as already mentioned, part of metatheoretical considerations about this knowledge in order to set, as its ideal goal, the thoroughgoing cognition of an object. In addition, the idea of a most real being is for Kant, as mentioned, an "analogue of a schema of sensibility", which could be conceived as an initial step of a Gödelian "idealization" of our intuition beyond the limits of sensibility (see footnote 3 above).

---

[9]See in [29] a reconstruction of Kant's critique of Leibniz's ontological argument on the ground of the texts available to Kant.

[10]"The analytic mark of possibility, which consists in the fact that mere positings (realities) do not generate a contradiction, of course, cannot be denied of this concept" [of a highest being] [16, B 630].

[11]"...the connection of all real properties *in a thing* is a synthesis" [16, B 630, our emphasis]. As Kant expresses himself, although the "*concept* of a highest being" has the "*analytic* mark of possibility", "we cannot judge a priori" about the *possibility* of a highest being, because it is *synthetic*, and "the mark of possibility of synthetic cognitions always has to be sought only in *experience*" ("to which...the object of an idea can never belong") [16, B 630, our emphasis].

[12]"...the famous Leibniz was far from having achieved what he flattered himself he had done, namely, gaining insight a priori into the possibility of such a sublime ideal being" [16, B 630, cf. B 329–330].

[13]This presupposition (*Voraussetzung*, cf. *suppositio relativa* [16, B 704]) should be distinguished from a hypothesis in an empirical theory. A hypothesis should be grounded on the real possibility of the object concerned [16, B 798].

3 Kant's concept of reality

The concept of "reality" is Kant's counterpart for (but not identical to) Descartes' and Leibniz's concept of "perfection". We will summarize Kant's view in the shape of several formally expressed principles, which in the next part of the paper will be included into a formalized theory as axioms or rules.[14]

Kant's introduction of the concept of reality is motivated by the metatheoretical principle of "thoroughgoing determination", which presupposes the "totality of predicates of things" (first-order one-place predicates) as already given. According to the principle, each object of a first-order theory should be determined with respect to each first-order predicate that could contribute to the determination of a thing:[15] "among all possible predicates of things, insofar as they are compared with their opposites, one must apply to it" [16, B 600]. In another formulation of the principle, a thing should be cognized with respect to "everything possible",[16] either affirmatively or negatively [16, B 601]. We see (1) that "all possible predicates of things" are arranged in possible pairs of contradictory predicates of things. Kant simply calls these predicates "possibilities", and further says (2) that in each such contradictory pair of predicates, according to their content (meaning), one predicate is "real" (meaning "reality", "thinghood", "being"), and the other one is negative (meaning "mere lack", "non-being in itself") [16, B 602–603].

The reference to *all* possible predicates of things ("contentual" predicates,[17] determining a thing), with the explicit quantification ("all") over predicates, prompts us to conceive first-order concepts as composing a domain on its own, above the first-order domain of possible sensible objects, and thus to extend our approach to a second-order setting. Such an approach will make it possible to propose a formal presentation of Kant's considerations on the "transcendental ideal of reason" and to more rigorously compare them with Gödel's ontotheology.[18]

---

[14]In a talk at a conference in Dubrovnik, May 2015, Paul Weingartner noticed in passing a connection between Kant's theory of the ideal of pure reason and Gödel's ontotheology.

[15]"The principle of thoroughgoing determination ... deals with the *content* and not merely the logical form. It is the principle of the *synthesis* of all predicates which are to make up the complete concept of a thing..." [16, B 600, our emphasis].

[16]Cf. "all possibility, which is supposed to contain a priori the data for the particular possibility of every thing" [16, B 601].

[17]See footnote 15 above.

[18]Besides, this approach could help us to shed some new light on the philosophical and historical background of Gödel's formalization of ontotheological questions, especially regarding Kant's critique of ontotheology. Possibly, Gödel wanted to meet Kant's objections and in this aspect to improve ontological proofs. According to Wang, Gödel "felt he needed to take Kant's critique of Leibniz seriously and find a way to meet Kant's objections to rationalism" [35, p. 164] (see [19, p. 149]).

3.1 The opposition of predicates

We proceed to Kant's examination of the opposition of real and negative predicates, and to present the results in the shape of formally expressed axioms about realities.

It is clear that, for Kant, only one of the two opposed predicates can be real. However, if we considered only logically opposed predicates, P and $\overline{P}$ (e.g., "mortal", "non-mortal", cf. [16, B 602]), we would not be able to decide which one of them is real, i.e., in itself signifies "being", and which one is negative, i.e., in itself signifies "mere lack" ("mere non-being", "non-being in itself"), because logical opposition does not pertain to the content of a concept taken in itself, but only to the relatedness of concepts to one another (see [16, B 602–603]). To stress the distinction between the affirmative content of a concept (as denoting "being", "thinghood") from the relative (logical) affirmation of a concept merely with respect to another concept, Kant calls the first kind of affirmation "transcendental" affirmation, and the corresponding negation (denoting "non-being") "transcendental" negation ("true negation", [16, B 604]):

> transcendental *affirmation* ... is called reality (thinghood), because through it alone, and only so far as it reaches, are objects Something (things); the opposed [transcendental] *negation*, on the contrary, signifies a mere *lack*, and where this alone is thought, the removal of all thing is represented. [16, B 602–603, our emphasis]

For example, "non-mortal" does not necessarily have negative meaning [16, B 602]. Besides, "darkness" [16, B 603] is logically positive but has negative content, "non-light", whereas "light", which is "non-darkness", is positive (real). Similarly, "poverty" [16, B 603] is with respect to its content negative, "non-prosperity", while "prosperity" (*Wohlstand*), which is "non-poverty", is real (positive). It follows that predicates should have a (positive or negative) *content* in order to qualify as real or negative. For example, as already mentioned, "existence", "being" is for Kant neither a real nor a ("contentually") negative but just a logical predicate, since it does not have any content by means of which it could enlarge a given concept [16, B 627, 623].

We can summarize the above description by saying that a predicate is real ($\mathscr{R}$) only if it and its opposite are contentual ($\mathscr{C}$, "content-related"), with the exclusion of existence (E) from the contentual predicates:

$$\mathscr{R}X \sqsupset_X \mathscr{C}X, \qquad \mathscr{R}X \sqsupset_X \mathscr{C}\overline{X}, \qquad \text{Cont1}$$

$$\neg \mathscr{C}E. \qquad \text{Cont2}$$

By $\sqsupset_X$ we denote a "hypothetical" proposition (a necessary conditional) that holds of each property X. Since contentuality, for Kant, pertains to predicates (concepts), it is natural to assume that the contentuality of predicates is preserved under their equivalence, if the equivalence includes a necessary connection between predicates. We express this by the following axiom:

$$(Tx \veebar_x \neg Ux) \sqsupset (\mathscr{C}T \sqsupset \mathscr{C}U), \qquad \text{Cont3}$$

where $\veebar_x$ stands for a necessary exclusive disjunction holding of each x. Finally, the exclusion condition between real and negative contentual properties can be formulated as follows: for each contentual predicate X it holds that either X or (exclusively) the negation $\overline{X}$ is real (this disjunction is expressed below as an abstract property by the λ-operator):

$$aX \, \mathscr{C}(\lambda X. \mathscr{R}X \veebar \mathscr{R}\overline{X}). \qquad \text{A1}$$

Operator aX is a universal quantifier applied to a subject-predicate sentence: in A1, $\mathscr{C}$ is the subject term and $(\lambda X. \mathscr{R}X \veebar \mathscr{R}\overline{X})$ the predicate term. A1 resembles Gödel's ontotheological axiom $\mathscr{P}X \veebar \mathscr{P}\overline{X}$, with the difference that we inserted the assumption of the contentuality, and replaced 'positive' with 'real' ($\mathscr{R}$), since Gödel does not exclude logical properties from the positive ones.[19]

3.2 The concept of the highest reality

Kant names "all of reality" (*omnitudo realitatis*) also the "highest reality" (*höchste Realität*), possession of which (*Allbesitz der Realität* [16, B 604]) completely determines one, "most real", being (*ens realissimum, das realste Wesen, das allerrealste Wesen*).[20] We can give the definition in Kant's words:

Definition 4 (Most real being (*ens realissimum*)).

> ... of all possible opposed predicates, one, namely that which belongs absolutely to being, is encountered in its determination.
>
> *[16, B 604]*[21]

---

[19]For Gödel's original ontological proof from 1970 and Scott's slight modifications (system GO), see [12] and [31]. For axiomatic reconstruction and semantics, see [3] and [32]. For a discussion on variants of the proof, see [14].

[20]Cf. [16, B 606, 607, 624, 631, 633, XXXII and elsewhere].

[21]"... und der Begriff eines *entis realissimi* ist der Begriff eines einzelnen Wesens, weil von allen möglichen entgegengesetzten Prädikaten eines, nämlich das, was zum Sein schlechthin gehört, in seiner Bestimmung angetroffen wird" [16, B 604].

Formally,

$$Rx =_{def} aX \mathscr{R}(\lambda X.Xx),$$

with R for 'is a most real being', which is defined by 'x possesses each real property'.
(Cf. Gödel's definition: $Gx =_{def} \forall X(\mathscr{P}X \to Xx)$, with Gx for 'x is God').

Furthermore, according to Kant, the highest being is completely determined by its concept, and is thus unchangeable and "eternal" [16, B 602, 604, 608], which can be expressed in terms of necessity:

$$Rx \sqsupset_x \Box Rx.$$

The corresponding proposition $\forall x(Gx \to \Box Gx)$ is provable in GO. In the formal system below, $Rx \sqsupset_x \Box Rx$ will be introduced as an axiom (A4).

Obviously, the "highest reality" is reality, and thus 'to be most real' is a real predicate. Hence, we add the axiom:

$$\mathscr{R}R. \hspace{8cm} \text{A2}$$

This corresponds to Gödel/Scott's axiom $\mathscr{P}G$ of the system GO, i.e., 'to be God' is a positive property. According to Gödel's original system, the intersection of "any number" of positive properties is positive. (Cf. [6, p. 148]).

3.3 Realities as consequents

Kant argues that the most real being (i.e., the "highest reality") should not itself be conceived as the totality ("all") of real predicates (with all realities as its "ingredients"), because that would mean that the highest reality is a sort of "aggregate" of "derived" beings. Obviously, if we take only a part of such an aggregate of all realities, with the lack of others, we obtain an individual being determined by the retained realities and the lack of others. Kant also adds that the conception of the most real being as a totality would mean that it would have sensible realities of the appearances as its "ingredients".

Thus, Kant proposes that the most real being should be conceived as the *ground* of all real predicates. The possibility of all things does not arise from the limitation of the most real being itself, but from the limitation of its consequents.[22] And negation

---

[22]"Rather, the highest reality would ground the possibility of all things as a *ground* and not as a *sum total*; and the manifoldness of the former rests not on the limitation of the original being itself, but on its complete consequence ...". "Vielmehr würde der Möglichkeit aller Dinge die höchste Realität als ein *Grund* und nicht als *Inbegriff* zum Grunde liegen und die Mannigfaltigkeit der ersteren nicht auf der Einschränkung des Urwesens selbst, sondern seiner vollständingen Folge beruhen..." [16, B 607].

would be grounded, in a sense, on the limitation of these consequents.[23] Kant's conception that the highest reality should be the *ground* of the totality of realities presupposes the idea of the closure of real predicates under the consequence relation, in the sense that each content-related predicate (V, of some real or existent subject T) possibly following from a real property (U, by definition possessed by the highest reality) of the subject T, is real ($\mathscr{R}$). Otherwise, V would contradict the real property non-V, which should also be possessed by the highest reality (A1, Definition 4). This can be expressed by a special rule:

$$\text{if (a) } \Gamma \vdash xTU \sqsupset_x xTV, \text{ (b) } \Gamma \vdash \mathscr{R}T \text{ or } \Gamma \vdash Tx \veebar \neg Ex, \\ \text{(c) } \Gamma \vdash \mathscr{R}U, \text{ and (d) } \Gamma \vdash \mathscr{C}V, \text{ then } \Gamma \vdash \mathscr{R}V. \quad \text{RC}$$

(Cf. axiom $(\mathscr{P}X \wedge \Box\forall y(Xy \to Yy)) \to \mathscr{P}Y$ of GO[24]). By the alternative condition $\Gamma \vdash Tx \veebar \neg Ex$, an existing "thing in itself" is hypothetically allowed as a subject of a real property (see footnote 3 above).

At first sight, RC leads to some strange conclusions. For example, if real sensible properties (like light, colour, red, warmth, gravity, weight, resistance, taste, etc.[25]) should not be ingredients of the most real being, then, according to the law of complete determination [16, B 599–600], the most real being would possess some negative predicates (that is, the negative counterparts of "light", "colour", "red", etc.). At the same time, to have such a real (although sensible) predicate should be a consequent of the definition of the concept of a most real being. Hence, the most real being would be a ground both of a real predicate and of its negation, that is, it would be impossible. However, let us note that R ("most real") could be a real property U of RC and be informally understood as some x's uncognized genuine ("constituent", see below Subsection 3.4) property, which is defined by the possibly consequent ("attributive", see Subsection 3.4) possession of real properties (Definition 4). Thus, a most real being might have some real property just in some mediate, secondary way (for example, due to the founding of a sensible world). Let us add that, possibly, a special, non-sensible ("spiritual", "noumenal") sense of a property should be distinguished from the sensible sense of the property, like in the above-mentioned examples of "light", "warmth", "prosperity" [16, B 603] (cf.

---

[23] Cf. "...no one can think a negation determinately without grounding it on the opposed affirmation" [16, B 603].

[24] See [13], where the corresponding axiom is expressed by means of "equality" of properties. For a complex form of hypothetical propositions as it is used in RC, cf. Kant's examples in his lectures on logic: 'If God is just, then the persistently Godless [evil] will be punished', i.e, [...then] 'he punishes the wicked' [18, Herschel 95–96, 98, 107, Vienna 932]. In *Prolegomena*: 'If a body is illuminated by the sun for long enough, then it becomes warm' [17, p. 312, §29] (see in [1]). Cf. also a discussion in [25, p. 103 ftn. 53].

[25] See, for instance, [16, B 211, 215, 216, 217, 603].

"light of knowledge" and "physical light").[26] It is only in the noumenal (spiritual) sense that they might be properly (constitutively) applicable to the most real being. Besides, predicates like "single", "simple", "all-sufficient" and "eternal", as genuinely applicable to the most real being [16, B 603], seem to be noumenal in their proper sense.

3.4 Necessity of realities

We come to the last part of an analysis of Kant's concept of "reality". A real property, for Kant, "belongs absolutely [*schlechthin*] to being" [16, B 604] – not only with respect to other concepts (A is B), but in itself and thus unchangeably.

> A transcendental negation ... signifies non-being in itself, and is opposed to transcendental affirmation, which is a Something, the concept of which in itself already expresses a being ... [16, B 602]

This "absoluteness" can be expressed by a universally affirmative proposition (quantifier aX) meaning that each property X which is real is necessarily real:

$$aX\,\mathscr{R}(\lambda X.\Box\mathscr{R}X). \qquad\qquad A3$$

(See Gödel's axiom $\mathscr{P}X \to \Box\mathscr{P}X$).

In connection with "necessity", we are led to the concept of *essence*, which has an important role in ontotheology. Kant distinguishes the "logical" and "real" essence of a thing, the first one pertaining to the concept of a thing (*esse conceptus*), and the second one to the "being" of a thing (*esse rei*) and conceived as consisting of "marks" (*Merkmale*) that necessarily belong to the thing. In addition, Kant distinguishes the constituents of an essence from the consequents of the essence (attributes) (see [15, refl. 2312, 2313, 2321–2323]). Generalizing the concept of consequence so as to include tautological cases (essence X following from X), we render Kant's concept of real essence by expressing that each content-related property necessarily possessed by x is a consequent of the essence of x:

Definition 5 (Essence).

$$Ess(X, x) =_{def} xX(\lambda x.aY\,(\mathscr{C}(\lambda Z.\Box Zx))(\lambda Y.Xy \sqsupset_y Yy))$$

('X is an essence of x iff x possesses X and each content-related necessary property Y of x follows from X').

---

[26] See [24].

This definition differs from Gödel's stronger definition, according to which *all* properties of a thing follow from its essence.[27] Kant's definition of essence is closer to Anderson's definition in a modified version of Gödel's ontological system: $\mathscr{E}(X, x) =_{def} \forall Y(\Box Yx \leftrightarrow \Box \forall y(Xy \rightarrow Yy))$ (see [2] and [14]).

4 Formal system

It is not hard to anticipate that, adopting the above-mentioned principles extracted from Kant's text, we can obtain a proof of the possibility of a most real being – in an analogy with Gödel's corresponding proof, although with some important differences. Since Kant's formal logic is not the same as what we today understand under "classical" or "standard" logic, we will present the proof of the possibility of a most real being in a formalization that will include some specific features of Kantian formal logic, like a basic logic of "categorical" (subject-predicate), "hypothetical", and exclusive disjunctive sentences. Conjunction directly applies only to predicates (for the "coordination" of "marks" in the content of a predicate). Universal subject-predicate sentences assume the possibility of their subjects (subject denotes a sort of possibly non-empty domain of predication). Hypothetical and disjunctive sentences are understood as strict conditional and strict exclusive disjunction, respectively. We disregard Kant's temporal (for him, non-logical) context of subject-predicate propositions (see [16, B 191–193]),[28] and follow his conception that, under some sufficient reason (here, a set of assumptions Γ), not only a predicate cannot contradict its subject (Kant's principle of contradiction [16, B 190]), but also predicates P and non-P cannot both belong to the same subject. Negation will be independent of the so called "existential import" of the subject-predicate sentences, and the principle of excluded middle thus upheld in full generality.[29]

We describe the language $\mathscr{L}\mathsf{CO}$ and define the system CO of a Kantian ontology.

The *vocabulary* of $\mathscr{L}\mathsf{CO}$: individual variables $x, y, z, x_1, \ldots$ (set $\mathcal{V}^1$), second-order variables: $X, Y, Z, X_1, \ldots$ (set $\mathcal{V}^2$), second-order constants $A, B, C, A_1, \ldots$ (set $\mathcal{A}$), and E ("existence"), third-order constants $\mathscr{R}$ ("real") and $\mathscr{C}$ ("contentual", "con-

---

[27] However, due to the modal "collapse" provable in GO, all properties of a thing x necessarily belong to x. See [31] for the proof, and [21] for the justification of the modal collapse from Gödel's philosophical viewpoint.

[28] For a tableau formalization, see [20]. The propositional part of this formalization is axiomatized in [27].

[29] For another formalization of Kant's formal logic, see [1]. For Kant's anticipations of modern logic, cf. [34]. See also [30] and [22]. A separate presentation of Kant's formal logic in first-order setting is in preparation.

tent-related"), operator symbols $\cdot, \lambda, \neg, a, \Box, \sqsupset_{(\alpha)}, \vee_{(\alpha)}$ (i.e., $\sqsupset$ and $\vee$ with or without $\alpha \in \mathcal{V}^1$ or $\alpha \in \mathcal{V}^2$ as a subscript), and parentheses.

S, P are used as metavariables for second-order subject and predicate terms, respectively, K for second-order constant terms, and T, U, V in general for second-order terms. Similarly, $\mathscr{S}, \mathscr{P}$, and $\mathscr{T}, \mathscr{U}$ are used for third-order terms. Also, α, β will stand for first-order or second-order variables, and γ for any term.

Definition 6 (Second-order term, T, third-order term, $\mathscr{T}$, formula, ϕ).

$$T ::= X \mid K \mid (T_1 \cdot T_2) \mid (\lambda x \phi), \quad X \in \mathcal{V}^2, K \in \mathcal{A} \cup \{E\}$$
$$\mathscr{T} ::= \mathscr{R} \mid \mathscr{C} \mid (\mathscr{T}_1 \cdot \mathscr{T}_2) \mid (\lambda X \phi)$$
$$\phi ::= Tx \mid \mathscr{T}T \mid ax\,SP \mid ax\,\neg SP \mid aX\,\mathscr{S}\mathscr{P} \mid aX\,\neg\mathscr{S}\mathscr{P} \mid \neg \phi \mid \Box \phi$$
$$\mid (\phi_1 \sqsupset_{(\alpha)} \phi_2) \mid (\phi_1 \vee_{(\alpha)} \phi_2)$$

Informally, we will usually use xTU and $T\mathscr{T}\mathscr{U}$ instead of $(T \cdot U)x$ and $(\mathscr{T} \cdot \mathscr{U})T$, respectively. Parentheses will be omitted if no ambiguity results. Instead of $\lambda \alpha \phi$, we will informally write $\lambda \alpha.\phi$.

For instance, Tx stands for informal 'x is T', $\neg Tx$ for 'x is not T', xTU for 'x is T and U', $ax\,SP$ (or $ax\,(\lambda x.Sx)(\lambda x.Px)$) for 'Each x which is S is P', and $ax\,\neg SP$ for 'No x which is S is P'.

$\overline{T}$ and $\overline{\mathscr{T}}$ abbreviate $\lambda x.\neg Tx$ and $\lambda X.\neg \mathscr{T}X$, respectively. $\lambda \alpha.\neg \phi$ will be occasionally expressed as $\overline{\lambda \alpha.\phi}$. Sometimes we will write $\bot$ for $xT\overline{T}$ or $X\mathscr{T}\overline{\mathscr{T}}$.

We note that SP is not a term (nor a subformula) of $ax\,SP$. Similarly, $\neg SP$ in $ax\,\neg SP$ is not a subformula (nor a term).

Variable α is *bound* by $a\alpha$, $\lambda \alpha$, $\sqsupset_\alpha$ and $\vee_\alpha$ ($a\alpha$ and nonobligatory α-subscript of $\sqsupset$ and $\vee$ are quantifiers). A term or a formula is closed if all variables occurring in it are bound, and is otherwise open. We denote by $\mathsf{free}(T)$, $\mathsf{free}(\mathscr{T})$, $\mathsf{free}(\phi)$ and $\mathsf{free}(\Gamma)$ the set of free variables occurring in T, $\mathscr{T}$, ϕ, and in the members of Γ, respectively.

By $\phi(\gamma)$ we will denote a formula ϕ possibly containing γ. $e(\gamma/\alpha)$ will denote the substitution of γ for α (if any) in the expression e. We also say that y is *substitutable* for x in an expression e iff y does not become bound if substituted for x in e. In a similar way, we say that T is substitutable for X in an expression e iff neither T nor a variable occuring in T becomes bound if substituted for X in e.

4.1 System CO

We formulate the rules and axioms of the system CO.

$\neg$I   If $\Gamma, \phi \vdash \bot$, then $\Gamma \vdash \neg \phi$
CI   if $\Gamma \vdash Tx$ and $\Gamma \vdash Ux$, then $\Gamma \vdash xTU$ if $\Gamma \vdash \mathscr{T}T$ and $\Gamma \vdash \mathscr{U}T$,

	then $\Gamma \vdash T\mathcal{TU}$	
CE	if $\Gamma \vdash xTU$, then $\Gamma \vdash Tx$	if $\Gamma \vdash T\mathcal{TU}$, then $\Gamma \vdash \mathcal{T}T$
	if $\Gamma \vdash xTU$, then $\Gamma \vdash Ux$	if $\Gamma \vdash T\mathcal{TU}$, then $\Gamma \vdash \mathcal{U}T$
Abs	$\Gamma \vdash \phi(y)$ iff $\Gamma \vdash (\lambda x.\phi)y$	$\Gamma \vdash \phi(T)$ iff $\Gamma \vdash (\lambda X.\phi)T$
aI	if $\Gamma \vdash Sz$ for some z, and $\Gamma \cup \{Sy\} \vdash Py$, then $\Gamma \vdash ax\, SP$	
	if $\Gamma \vdash Sz$ for some z, and $\Gamma \cup \{Sy\} \vdash \neg Py$, then $\Gamma \vdash ax\, \neg SP$,	
	where $y \notin \mathsf{free}(\Gamma, ax\, SP/ax\, \neg SP)$	
aE	if $\Gamma \vdash ax\, SP$, then (i) if $\Gamma \cup \{Sz\} \vdash \phi$ then $\Gamma \vdash \phi$ and (ii) if $\Gamma \vdash Sy$	
	then $\Gamma \vdash Py$	
	if $\Gamma \vdash ax\, \neg SP$, then (i) if $\Gamma \cup \{Sz\} \vdash \phi$ then $\Gamma \vdash \phi$ and (ii) if $\Gamma \vdash Sy$	
	then $\Gamma \vdash \neg Py$,	
	where $z \notin \mathsf{free}(\Gamma, ax\, SP/ax\, \neg SP, \phi)$, y is substitutable for x in	
	$SP/\neg SP$	
E	$\Gamma \vdash ax\, EE$.	

The following are the second-order rules for universal sentences:

aI2	if $\Gamma \vdash \mathcal{S}T$ for some T, and $\Gamma \cup \{\mathcal{S}Y\} \vdash \mathcal{P}Y$, then $\Gamma \vdash aX\, \mathcal{SP}$
	if $\Gamma \vdash \mathcal{S}T$ for some T, and $\Gamma \cup \{\mathcal{S}Y\} \vdash \neg \mathcal{P}Y$, then $\Gamma \vdash aX\, \neg\mathcal{SP}$,
	where $Y \notin \mathsf{free}(\Gamma, aX\, \mathcal{SP}/aX\, \neg\mathcal{SP})$
aE2	if $\Gamma \vdash aX\, \mathcal{SP}$, then (i) if $\Gamma \cup \{\mathcal{S}Z\} \vdash \phi$ then $\Gamma \vdash \phi$ and (ii) if $\Gamma \vdash$
	$\mathcal{S}U$ then $\Gamma \vdash \mathcal{P}U$
	if $\Gamma \vdash aX\, \neg\mathcal{SP}$, then (i) if $\Gamma \cup \{\mathcal{S}Z\} \vdash \phi$ then $\Gamma \vdash \phi$ and (ii) if $\Gamma \vdash$
	$\mathcal{S}U$ then $\Gamma \vdash \neg\mathcal{P}U$,
	where $Z \notin \mathsf{free}(\Gamma, aX\,\mathcal{SP}/aX\,\neg\mathcal{SP}, \phi)$, U is substitutable for X in
	$\mathcal{SP}/\neg\mathcal{SP}$.

Let $\sqrt{\Gamma}$ abbreviate $\{\phi \mid \Gamma \vdash \Box\phi \text{ or } \Gamma \vdash \phi = \psi \sqsupset_{(\alpha)} \chi \text{ or } \Gamma \vdash \phi = \psi \veebar_{(\alpha)} \chi\}$. We give rules for $\Box, \sqsupset_{(\alpha)}$ and $\veebar_{(\alpha)}$:

S4	modal propositional rules and axioms of system S4: K- and 4-Reiteration into a $\Box$-subproof, $\Box$I, strict Axiom T ($\Box\phi \sqsupset \phi$); the reiteration of $\phi(\alpha) \sqsupset_{(\alpha)} \psi(\alpha)$ and $\phi(\alpha) \veebar_{(\alpha)} \psi(\alpha)$ into a $\Box$-subproof
$\sqsupset$I	if $\sqrt{\Gamma}, \phi \vdash \psi$ then $\Gamma \vdash \phi(\alpha/\beta) \sqsupset_{(\alpha)} \psi(\alpha/\beta)$, where, if $\sqsupset_{(\alpha)} = \sqsupset_\alpha$ then $\beta \notin \mathsf{free}(\Gamma)$, and if $\sqsupset_{(\alpha)} = \sqsupset$ then $\beta = \alpha$
MP	if $\Gamma \vdash \phi \sqsupset_{(\alpha)} \psi$, then if $\Gamma \vdash \phi(\gamma/\alpha)$ then $\Gamma \vdash \psi(\gamma/\alpha)$, where γ is substitutable for α in $\phi \sqsupset \psi$ and, if $\sqsupset_{(\alpha)} = \sqsupset$, $\gamma = \alpha$
$\veebar$I	if $\sqrt{\Gamma}, \phi \vdash \neg\psi$ and $\sqrt{\Gamma}, \neg\phi \vdash \psi$, then $\Gamma \vdash \phi(\alpha/\beta) \veebar_{(\alpha)} \psi(\alpha/\beta)$, where, if $\sqsupset_{(\alpha)} = \sqsupset_\alpha$ then $\beta \notin \mathsf{free}(\Gamma)$, and if $\sqsupset_{(\alpha)} = \sqsupset$ then $\beta = \alpha$

MPT   if $\Gamma \vdash \phi \veebar_{(\alpha)} \psi$, then if $\Gamma \vdash \phi(\gamma/\alpha)$ then $\Gamma \vdash \neg\psi(\gamma/\alpha)$, where γ is substitutable for α in $\phi \veebar \psi$ and, if $\beth_{(\alpha)}{=}\beth$, $\gamma = \alpha$

MTP   if $\Gamma \vdash \phi \veebar_{(\alpha)} \psi$, then if $\Gamma \vdash \neg\phi(\gamma/\alpha)$ then $\Gamma \vdash \psi(\gamma/\alpha)$, where γ is substitutable for α in $\phi \veebar \psi$ and, if $\beth_{(\alpha)}{=}\beth$, $\gamma = \alpha$

REM $\Gamma \vdash \phi(\alpha) \veebar_{(\alpha)} \neg\phi(\alpha)$.

Reiteration rule: if $\Gamma \vdash \phi$, then $\Gamma, \Delta \vdash \phi$.   *Assumption* rule: $\Gamma, \phi \vdash \phi$.

As already mentioned, Kant requires that the predicate of a proposition should not contradict its subject (principle of contradiction [16, B 190–193]). Consequently, to take a first-order example, neither T nor $\overline{T}$ hold of a self-contradictory subject ($T\overline{T}$) [17, p. 341, §52b]. The requirement for a subject term in the aI rule ($\Gamma \vdash Sz$) roughly reflects Kant's general statement that both the affirmation as well as the negation of a predicate are "incorrect" if they have as "their ground an impossible concept of the object" (*non entis nulla sunt praedicata* [16, B 820–821]). Kant has in mind the non-contradictoriness of the concept S, which is strengthened in aI up to the derivability of S's application to a possible x (see also the note immediately after the proof of Theorem 1 below).

Let us remark that the square of opposition holds between $axSP$, $ax\neg SP$, $\neg axSP$ and $\neg ax\neg SP$ if the condition of the derivability of Sz (for some z) is *assumed* to be fulfilled for each of these sentences, that is, if all the sentences of the square are viewed as candidate consequents of a set Γ implying Sz. At the same time, if the fulfilment of the condition of Sz is not assumed but only added as *required* for each of the four forms, REM does not hold between traditional "contradictories" (between ax- and $\neg ax$-sentences). We note that in this case the forms $\neg axSP$ and $\neg ax\neg SP$ should be accompanied by the additional condition of Sz, for example, by the transformation into the following formulas: $azS(\lambda y.\neg ax\neg SP)$ for a "particular affirmative" sentence (i), and $azS(\lambda y.\neg axSP)$ for a "particular negative" sentence (o). Thus, evidently, both "subcontraries" can also be denied.

Proposition 1 (Derived rules $\neg aE$ and $\neg aE2$). *(1) If $\Gamma \vdash \neg axSP$, then if (a) $\Gamma \vdash Sz$ for some z and (b) $\Gamma \cup \{Sy, \neg Py\} \vdash \bot$ ($y \notin \text{free}(\Gamma, \neg axSP)$) then $\Gamma \vdash \phi$ ($\neg aE$). (2) If $\Gamma \vdash \neg aXSP$, then if (a) $\Gamma \vdash \mathscr{S}T$ for some T and (b) $\Gamma \cup \{\mathscr{S}Y, \neg\mathscr{P}Y\} \vdash \bot$ ($Y \notin \text{free}(\Gamma, \neg aX\mathscr{S}\mathscr{P})$) then $\Gamma \vdash \phi$ ($\neg aE2$).*

Proof. ($\neg aE$) The condition (a) is the first conjunct of the condition of the rule aI, and from (b) the second conjunct of the condition of aI follows. Thus, $axSP$ is provable from Γ. From this and from the assumption ($\Gamma \vdash \neg axSP$), $x(\lambda x.axSP)\overline{(\lambda x.axSP)}$ follows, and thus $\Gamma \vdash \phi$. ($\neg aE2$) The proof is similar to that for $\neg aE$. □

We now add specific *ontotheological* definitions and axioms.

Definition 7 (Most real being, R (= Definition 4)). $Rx =_{def} aX\mathscr{R}(\lambda X.Xx)$

Definition 8 (Essence, Ess (= Definition 5)). $Ess(X,x) =_{def} xX(\lambda x.aY(\mathscr{C}(\lambda Z.\Box Zx))(\lambda Y.Xy \sqsupset_y Yy))$

Ontotheological *axioms* and the *rule* RC:

A1	$aX\mathscr{C}(\lambda X.\mathscr{R}X \veebar \mathscr{R}\overline{X})$
A2	$\mathscr{R}R$
A3	$aX\mathscr{R}(\lambda X.\Box\mathscr{R}X)$
A4	$Rx \sqsupset_x \Box Rx$
Cont1	$\mathscr{R}X \sqsupset_X \mathscr{C}X \quad \mathscr{R}X \sqsupset_X \mathscr{C}\overline{X}$
Cont2	$\neg\mathscr{C}E$
Cont3	$(Tx \veebar_x \neg Ux) \sqsupset (\mathscr{C}T \sqsupset \mathscr{C}U)$
RC	if (a) $\Gamma \vdash xTU \sqsupset_x xTV$, (b) $\Gamma \vdash \mathscr{R}T$ or $\Gamma \vdash Tx \veebar \neg Ex$, (c) $\Gamma \vdash \mathscr{R}U$, and (d) $\Gamma \vdash \mathscr{C}V$, then $\Gamma \vdash \mathscr{R}V$.

Definition 9 (Inconsistency and consistency). *A set Γ of sentences of $\mathscr{L}$CO is inconsistent iff $\Gamma \vdash \bot$. Otherwise, Γ is consistent.*

5 The possibility of the most real being

Proposition 2. $\neg aX\neg\mathscr{C}\mathscr{R}$ *(Since there are content-related properties, the proposition means 'Some content-related property is real').*

Proof. See axioms A2 and Cont1, which also show that there are content-related properties. In addition, it is impossible that no content-related property is real since otherwise, for each content-related property T, neither T nor its complement $\overline{T}$ would be real, contrary to A1. □

Gödel introduces the exemplifying proposition about the positivity of self-identity: $\mathscr{P}(\lambda x.x = x)$. However, if preferred, the use of self-identity can be left out, see [31, p. 120].

Theorem 1. $aX\mathscr{R}(\lambda X.\neg\Box ax\neg EX)$ *(Since there always are existing objects, the meaning is 'Each real property is possibly instantiated by an existing object').*

Cf. the theorem $\mathscr{P}X \to \Diamond\exists x\,Xx$ of the system GO.

Proof.

1	$\mathscr{R}R$	A2
2	$\mathscr{R}Y$	assumption

3	$\Box ax\neg EY$	assumption
4	yER	$\Box$-assumption
5	$ax\neg EY$	from 3 K-Reiteration
6	Ey	from 4 CE
7	$\neg Yy$	from 5, 6 aE
8	$\overline{Y}y$	from 7 Abs
9	$yE\overline{Y}$	from 6, 8 CI
10	$yER \sqsupset_y yE\overline{Y}$	from 4–9 $\sqsupset$I, assumption 4 and $\Box$ discharged
11	$\mathscr{C}Y$	from 2 Cont1, MP
12	$\mathscr{C}\overline{Y}$	from 2 Cont1, MP
13	$\mathscr{R}\overline{Y}$	from 10, 12 RC
14	$\overline{\mathscr{R}}\,\overline{Y}$	from 2, 11 A1, aE2, MPT
15	$\overline{Y}\mathscr{R}\overline{\mathscr{R}}$	from 13, 14 CI
16	$\neg\Box ax\neg EY$	from 3–15 $\neg$I, assumption 3 discharged
17	$aX\mathscr{R}(\lambda X.\neg\Box ax\neg EX)$	from 1, 2–16 aI2, assumption 2 discharged

$\Box$

We note that the possibility of the subject E of $\neg\Box ax\neg EX$ follows from Axiom E. In general, assume $\neg Ex$ as a valid scheme. Then, from the assumption Ez, $zE\overline{E}$ and thus $yE\overline{E}$ follow, which contradicts the axiom $axEE$ (see aE). Hence $\neg Ex$ is not a valid scheme, and there should always be some x such that Ex holds. This is also relevant for the next proposition.

Proposition 3. $\neg\Box ax\neg ER$ *('Possibly, a most real being exists')*.

Cf. Gödel's proposition $\Diamond\exists xGx$.

Proof. From Theorem 1 and A2, $(\lambda X.\neg\Box ax\neg EX)R$ follows by aE2(ii), and gives by Abs the proposition. $\Box$

Proposition 4. $\neg\Box ax\neg E(\lambda x.\Box Rx)$ *('Possibly, a necessary most real being exists')*.

Proof. We give a shortened overview of the proof. Assume $\Box ax\neg E(\lambda x.\Box Rx)$ (1). In a $\Box$-subproof assume $\neg ax\neg ER$ (2). In (3) derive $ax\neg E(\lambda x.\Box Rx)$ from (1) by K-Reiteration. For some z, assume Ez (4) (for aE, cf. (3)). Assume Ey and Ry (5), with $y \notin \text{free}(\Gamma, \neg ax\neg ER)$ (Γ is the set of current assumptions). $\Box Ry$ (6) follows from (5) and A4. However, from (3), $\neg\Box Ry$ (7) is derivable, and hence $\bot$ (8). From (2), (4), and (5–8) $\bot$ (9) follows (derived rule $\neg a$E). From (3) and (4–9) $\bot$ (10) is derivable (aE(i)). From (2-10) $ax\neg ER$ (11) follows ($\neg$I, REM) and in (12) by $\Box$I, we obtain $\Box ax\neg ER$, contradicting Proposition 3, with the derivability of $\bot$. Hence the proposition follows by the negation of (1). $\Box$

An interesting derivable proposition is that if something with a positive property necessarily exists, then something else, too, possibly exists.

Proposition 5. $\mathscr{R}T \sqsupset (\Box ax\,TE \sqsupset \neg\Box ax\,ET)$.

Proof. It can be seen that $\mathscr{R}T$ and $\Box ax\,TE$, with the assumption $\Box ax\,ET$, imply $Ty \veebar_y \neg Ey$. Hence, $\mathscr{C}E$ follows (from $\mathscr{R}T$, Cont1, and Cont3), in contradiction to Cont2. Thus, the assumption $\Box ax\,ET$ should be negated. $\square$

An analogous proposition holds for $\overline{T}$ instead of T.

Proposition 6. $Rx \sqsupset_x aX(\mathscr{C}(\lambda X.Xx))\mathscr{R}$ (*'If something is most real, then all its content-related properties are real'*).

Cf. proposition $Gx \to (Xx \to \mathscr{P}X)$ in GO.

Proof.

1	Rx	$\Box$-assumption
2	$R\mathscr{C}(\lambda X.Xx)$	A2, Cont1, Abs from 1, CI
3	$X\mathscr{C}(\lambda X.Xx)$	assumption
4	$\neg\mathscr{R}X$	assumption
5	$\mathscr{R}\overline{X}$	from 3, 4 A1, MTP
6	$xX\overline{X}$	from 3, 5, 1 Definition 7, CI
7	$\mathscr{R}X$	from 4–6 $\neg$I, REM, assumption 4 discharged
8	$aX(\mathscr{C}(\lambda X.Xx))\mathscr{R}$	2, 3–7 aI2, assumption 3 discharged
9	$Rx \sqsupset_x aX(\mathscr{C}(\lambda X.Xx))\mathscr{R}$	from 1–8 $\sqsupset$I, assumption 1 and $\Box$ discharged

$\square$

Proposition 7. $\forall X\,\mathscr{R}(\lambda X.Rx \sqsupset_x Xx)$ (*'A most real being is a ground of all real properties'*).

Proof. Follows from A2, and from Definition 7 and A3. $\square$

Theorem 2. $Rx \sqsupset_x Ess(R, x)$ (*'If something is most real, then it is its essence to be most real'*).

Cf. Gödel's theorem $Gx \to Ess(G, x)$.

Proof. Assume Rx in a $\Box$-subproof. We should prove Rx (trivially) and $aY(\mathscr{C}(\lambda Z.\Box Zx))(\lambda Y.Ry \sqsupset_y Yy)$ (cf. Definition 8). According to aI2, we should first prove that the subject term holds for some T substituted for Y. To this end we can chose R itself, which is obviously content-related ($\mathscr{C}$, A2, Cont1), and $\Box Rx$ follows from the assumption (A4). Then, according to aI2, we prove $Ry \sqsupset_y Yy$ from the assumption

$Y\mathscr{C}(\lambda Z.\Box Zx)$, for an arbitrary Y. From $Y\mathscr{C}(\lambda Z.\Box Zx)$ propositions $\mathscr{C}Y$ and $\Box Yx$ follow, and thus $\mathscr{R}Y$ (T, Proposition 6) and $\Box\mathscr{R}Y$ (A3, aE). According to Definition 7 of R, we obtain $Ry \sqsupset_y Yy$. □

In Kant's critical philosophy, space and time are not determinations of things in themselves, but are due to the *form* of our sensible intuition, which is the way the objects can be given to us [16, B 323–324, 607].[30] Thus, we define *conceptual identity* of things, in accordance with which the corresponding manifoldness consists in the difference between things regarding their real properties [16, B 606, 322–323].

Definition 10 (Conceptual identity). $x \equiv y =_{def} aY\mathscr{R}((\lambda X.Xx \sqsupset Xy)(\lambda X.Xy \sqsupset Xx))$ *('The conceptual identity of x and y means that a real property belongs to x iff this property belongs to y').*

On the non-Kantian presupposition that each thing has its corresponding individual concept (real property), the definiens could be shortened to $aY\mathscr{R}(\lambda X.Xx \sqsupset Xy)$.

Proposition 8. $Rx \sqsupset_x (Ry \sqsupset_y x \equiv y)$ *('If x is most real, then, if y is most real, x is conceptually identical to y').*

In Gödel's system, $Gx \to (Gy \to x = y)$ is provable for object identity (=).

Proof. Follows from definitions 10 and 7, with the use of axioms A2, A3, and A4 (x and y share all real properties). □

In CO, the preservation of "reality" can be proved under the equivalence of concepts.

Proposition 9 (Reality and equivalent concepts). $(Tx \veebar_x Ux) \sqsupset (\mathscr{R}T \sqsupset \mathscr{R}U)$

Proof. Assume $Tx \veebar_x Ux$ (1) in a □-subproof. In a new, subordinated □-subproof assume $\mathscr{R}T$ (2). We derive $xTT \sqsupset_x xTU$ (3) (from 1) and $\mathscr{C}U$ (4) (from 1 and 2, by Cont1 and Cont3). By RC, we deduce $\mathscr{R}U$ (5) (from 3, 2, 4). Discharging the inner □-subproof, we derive $\mathscr{R}T \sqsupset \mathscr{R}U$ (6), and with the first □-subproof discharged, the proposition follows (7). □

The corresponding proposition on positivity ($\mathscr{P}$), $\Box\forall x(Tx \leftrightarrow Ux) \to (\mathscr{P}T \to \mathscr{P}U)$, is easily provable in GO).

---

[30]Regardless of a content-related difference (with respect to the concept of a thing), things can be additionally distinguished, for example, according to the difference in the space-time position, space-time direction, or quality of sensation (enjoyment, pain) [16, B 319–321, 329–330].

6 Frames and models

We use semantics with varying domains. However, quantifiers are possibilistic and range over the objects of a frame (model) and over concepts (functions). Quantifier $a\alpha$ is accompanied with the explicit restriction to the objects satisfying the subject term.

Definition 11 (Frame). *Frame $\mathfrak{F}$ is a sextuple $\langle W, R, D, D(1), q, I \rangle$, where $W \neq \varnothing$ (non-empty set of "worlds"), $R \subseteq W \times W$ (reflexive, transitive), $D \neq \varnothing$ (non-empty set of objects), $\varnothing \neq D(1) \subseteq \wp D^W$ (non-empty set of concepts), $q \in D(1)$ with $q(w) \neq \varnothing$ (world-relative domains), and I is an interpretation function such that $I(K \in \mathcal{A}) \in D(1)$, $I(E) = q$, $I(\mathscr{C}, w) \subseteq D(1) \setminus \{q\}$, $I(\mathscr{R}, w) \subseteq I(\mathscr{C}, w)$, and $I(\lambda x.\phi) \in D(1)$.*

Definition 12 (Variable assignment). *Variable assignment v is a mapping from $\mathcal{V}^1$ to D, and from $\mathcal{V}^2$ to $D(1)$.*

Definition 13 (Denotation of a term).

1. $[\![x]\!]_v^{\mathfrak{F},w} = [\![x]\!]_v^{\mathfrak{F}} = v(x)$, $[\![X]\!]_v^{\mathfrak{F},w} = v(X)(w)$, $[\![X]\!]_v^{\mathfrak{F}} = v(X)$,
2. $[\![K]\!]_v^{\mathfrak{F},w} = I(K, w)$, $[\![K]\!]_v^{\mathfrak{F}} = I(K)$, where $K \in \mathcal{A} \cup \{E\}$,
3. $[\![\mathscr{C}]\!]_v^{\mathfrak{F},w} = I(\mathscr{C}, w)$, $\quad [\![\mathscr{R}]\!]_v^{\mathfrak{F},w} = I(\mathscr{R}, w)$,
4. $[\![\lambda x.\phi]\!]_v^{\mathfrak{F},w} = [\![\lambda x.\phi]\!]_v^{\mathfrak{F}}(w)$,
5. $[\![\lambda X.\phi]\!]_v^{\mathfrak{F},w} \in \wp D(1)$.

Definition 14 (Satisfaction of a formula at a world).

1. $\mathfrak{F}, w \models_v Tx$ iff $v(x) \in [\![T]\!]_v^{\mathfrak{F},w}$,
   $\mathfrak{F}, w \models_v \mathscr{T}T$ iff $[\![T]\!]_v^{\mathfrak{F}} \in [\![\mathscr{T}]\!]_v^{\mathfrak{F},w}$,
2. $\mathfrak{F}, w \models_v \neg\phi$ iff $\mathfrak{F}, w \not\models_v \phi$,
3. $\mathfrak{F}, w \models_v xTU$ iff $\mathfrak{F}, w \models_v Tx$ and $\mathfrak{F}, w \models_v Ux$,
   $\mathfrak{F}, w \models_v T\mathscr{T}\mathscr{U}$ iff $\mathfrak{F}, w \models_v \mathscr{T}T$ and $\mathfrak{F}, w \models_v \mathscr{U}T$,
4. $\mathfrak{F}, w \models_v ax\,SP$ iff $[\![S]\!]_v^{\mathfrak{F},w} \subseteq [\![P]\!]_v^{\mathfrak{F},w}$ with $[\![S]\!]_v^{\mathfrak{F},w} \neq \varnothing$,
   $\mathfrak{F}, w \models_v ax\,\neg SP$ iff $[\![S]\!]_v^{\mathfrak{F},w} \cap [\![P]\!]_v^{\mathfrak{F},w} = \varnothing$ with $[\![S]\!]_v^{\mathfrak{F},w} \neq \varnothing$,
   $\mathfrak{F}, w \models_v aX\,\mathscr{S}\mathscr{P}$ iff $[\![\mathscr{S}]\!]_v^{\mathfrak{F},w} \subseteq [\![\mathscr{P}]\!]_v^{\mathfrak{F},w}$ with $[\![\mathscr{S}]\!]_v^{\mathfrak{F},w} \neq \varnothing$,
   $\mathfrak{F}, w \models_v aX\,\neg\mathscr{S}\mathscr{P}$ iff $[\![\mathscr{S}]\!]_v^{\mathfrak{F},w} \cap [\![\mathscr{P}]\!]_v^{\mathfrak{F},w} = \varnothing$ with $[\![\mathscr{S}]\!]_v^{\mathfrak{F},w} \neq \varnothing$,
5. $\mathfrak{F}, w \models_v \Box\phi$ iff for each w' with wRw', $\mathfrak{F}, w' \models_v \phi$,

6. $\mathfrak{F}, w \models_v \phi \sqsupset_{(\alpha)} \psi$ iff for each w' with wRw', for each d in the domain for α, if $\mathfrak{F}, w' \models_{v[d/\alpha]} \phi$, then $\mathfrak{F}, w' \models_{v[d/\alpha]} \psi$ (if $\sqsupset_{(\alpha)} = \sqsupset$, then $v[d/\alpha] = v^{31}$),

7. $\mathfrak{F}, w \models_v \phi \veebar_{(\alpha)} \psi$ iff for each w' with wRw', for each d in the domain for α, either $\mathfrak{F}, w' \models_{v[d/\alpha]} \phi$ and $\mathfrak{F}, w' \models_{v[d/\alpha]} \neg\psi$, or $\mathfrak{F}, w' \models_{v[d/\alpha]} \psi$ and $\mathfrak{F}, w' \models_{v[d/\alpha]} \neg\phi$ (if $\sqsupset_{(\alpha)} = \sqsupset$, then $v[d/\alpha] = v$, see footnote 31).

Now, to make sure that each $\lambda\alpha.\phi$ abstract has the intended meaning (semantic value), corresponding to the meaning of ϕ, and to give the intended meanings to $\mathscr{C}$ and $\mathscr{R}$, we restrict frames to models, where the meanings of $\lambda\alpha.\phi$, $\mathscr{C}$ and $\mathscr{R}$ are defined by means of the satisfaction of formulas.

Definition 15 (Model). *Model $\mathfrak{M}$ is a frame $\mathfrak{F}$ where*

1. $[\![\lambda x.\phi]\!]_v^{\mathfrak{F},w} = \{d \in D \mid \mathfrak{F}, w \models_{v[d/x]} \phi\}$,
   $[\![\lambda X.\phi]\!]_v^{\mathfrak{F},w} = \{d \in D(1) \mid \mathfrak{F}, w \models_{v[d/X]} \phi\}$,

2. $\{[\![T]\!]_v^{\mathfrak{F}} \mid T = U \text{ or } T = \overline{U} \text{ and } [\![U]\!]_v^{\mathfrak{F}} \in [\![\mathscr{R}]\!]_v^{\mathfrak{F},w}\} \subseteq [\![\mathscr{C}]\!]_v^{\mathfrak{F},w}$, $\{q\} \notin [\![\mathscr{C}]\!]_v^{\mathfrak{F},w}$,

3. $[\![\mathscr{R}]\!]_v^{\mathfrak{F},w} \in \wp D(1)$ and fulfils the following conditions:[32]

    (a) if $[\![T]\!]_v^{\mathfrak{F}} \in [\![\mathscr{C}]\!]_v^{\mathfrak{F},w}$, then $[\![T]\!]_v^{\mathfrak{F}} \in [\![\mathscr{R}]\!]_v^{\mathfrak{F},w}$ iff $[\![\overline{T}]\!]_v^{\mathfrak{F}} \notin [\![\mathscr{R}]\!]_v^{\mathfrak{F},w}$,

    (b) if $[\![T]\!]_v^{\mathfrak{F},w} = \bigcap\{[\![U]\!]_v^{\mathfrak{F},w} \mid [\![U]\!]_v^{\mathfrak{F}} \in [\![\mathscr{R}]\!]_v^{\mathfrak{F},w}\}$, then $[\![T]\!]_v^{\mathfrak{F}} \in [\![\mathscr{R}]\!]_v^{\mathfrak{F},w}$ and for $w'Rw''$, $[\![T]\!]_v^{\mathfrak{F},w'} \subseteq [\![T]\!]_v^{\mathfrak{F},w''}$,

    (c) if $[\![T]\!]_v^{\mathfrak{F}} \in [\![\mathscr{R}]\!]_v^{\mathfrak{F},w}$, then $\forall w'$ with wRw', $[\![T]\!]_v^{\mathfrak{F}} \in [\![\mathscr{R}]\!]_v^{\mathfrak{F},w'}$,

    (d) if (a) $\forall w'$ with wRw', $[\![(TU)]\!]_v^{\mathfrak{F},w'} \subseteq [\![(TV)]\!]_v^{\mathfrak{F},w'}$, (b) either $[\![T]\!]_v^{\mathfrak{F}} \in [\![\mathscr{R}]\!]_v^{\mathfrak{F},w}$ or $\forall w'$ with wRw', $[\![T]\!]_v^{\mathfrak{F},w'} = [\![E]\!]_v^{\mathfrak{F},w'}$, (c) $[\![U]\!]_v^{\mathfrak{F}} \in [\![\mathscr{R}]\!]_v^{\mathfrak{F},w}$, and (d) $[\![V]\!]_v^{\mathfrak{F}} \in [\![\mathscr{C}]\!]_v^{\mathfrak{F},w}$, then $[\![V]\!]_v^{\mathfrak{F}} \in [\![\mathscr{R}]\!]_v^{\mathfrak{F},w}$.

Definition 16 (Consequence). $\Gamma \models \phi$ *iff for each model $\mathfrak{M}$, world w and assignment v, if $\mathfrak{M}, w \models_v \psi$ for each $\psi \in \Gamma$, then $\mathfrak{M}, w \models_v \phi$.*

Example 1 (A possible highest being). *We describe a very simple model $\mathfrak{M}^P$, with two worlds and two objects. The highest being exists only in one of them, while in the other one (say, "empirical") it is just a possibility. $W = \{w_1, w_2\}$, $R = \{\langle w_1, w_1\rangle, \langle w_1, w_2\rangle, \langle w_2, w_2\rangle\}$, $D = \{d, g\}$, $D(1) \subseteq \wp D^W$, $q(w_1) = \{d\}$, $q(w_2) = \{g\}$, and for both worlds w, $[\![R]\!]_v^{\mathfrak{F},w} = \{g\}$. Accordingly, $\mathfrak{M}^P, w_1 \models ax\neg ER$, but $\mathfrak{M}^P, w_1 \models \neg\Box ax\neg ER$ and $\mathfrak{M}^P, w_1 \models \neg\Box ax\neg E(\lambda x.\Box Rx)$. Also, $\mathfrak{M}^P, w_2 \models \neg\Box ax\neg ER$, but $\mathfrak{M}^P, w_2 \models \Box axER$ and $\mathfrak{M}^P, w_2 \models \Box axE(\lambda x.\Box Rx)$.*[33]

---

[31] The insertion "for each d in the domain for α" becomes redundant.

[32] Compare the conditions for $I(\mathscr{P}, w)$ in [3, 23].

[33] See, for example, [4] and [33] for different ways to supplement a modal **S4** basis in order to obtain the necessary existence of a highest being as a validity.

6.1 Adequacy

We give an outline and main details of the soundness and completeness proofs for the logic CO.

Theorem 3 (Soundness). *If $\Gamma \vdash \phi$ then $\Gamma \models \phi$.*

Proof. Let us focus on some characteristic cases. (a) ($\neg$I) The truth of $xT\overline{T}$ ($\bot$) implies the truth of Tx and $\overline{T}x$, which is impossible under the same model and variable assignment (Definition 14, case 2). (b) (aI) Assume the antecedent of aI (for $axSP$), but let, under inductive hypothesis, $\Gamma \not\models axSP$. Hence, there should be some context $\mathfrak{M}, w, v$ which satisfies Γ and in which either $[\![S]\!]_v^{\mathfrak{M},w} = \varnothing$, contradicting the hypothesis (with respect to the first conjunct of the antecedent of aI), or for some d, $d \in [\![S]\!]_v^{\mathfrak{M},w}$ but $d \notin [\![P]\!]_v^{\mathfrak{M},w}$. In the latter case, there is a variant assignment $v[d/y]$ satisfying Sy but not Py ($y \notin \mathsf{free}(axSP)$). This contradicts the hypothesis with respect to the second conjunct of the antecedent of aI since the value of y leaves Γ satisfied ($y \notin \mathsf{free}(\Gamma)$). (c) For ($a$E), let us just mention that, under the proviso for z, from the semantic non-emptiness of S (Definition 14, case 4), it follows that $\Gamma \cup \{Sz\} \models \phi$ implies $\Gamma \models \phi$ (semantic counterpart of the conjunct (i) in aE). For the proof, assume that $\Gamma \cup \{Sz\} \models \phi$ and $\Gamma \not\models \phi$. Hence, there is a context $\mathfrak{M}, w, v$ that satisfies Γ but does not satisfy ϕ, and (because of the non-emptiness of $[\![S]\!]_v^{\mathfrak{M},w}$) there is a z-variant v' of v that satisfies Sz (without disturbing the (non-)satisfaction of Γ and ϕ because of the proviso for z in aE). Thus, $\mathfrak{M}, w, v'$ contradicts the assumption that $\Gamma \cup \{Sz\} \models \phi$. (d) S4 modal rules hold because of the reflexive and transitive accessibility of worlds in a model. (e) Rules for $\sqsupset_{(\alpha)}$ and $\veebar_{(\alpha)}$ are semantically obvious since they describe necessary (universally quantified) conditional and necessary (universally quantified) exclusive disjunction. MP, MPT, and MTP hold because of reflexivity. (f) Let us take RC as an example of the specific ontotheological part of the system. Assume $\Gamma \vdash xTU \sqsupset_x xTV$, $\Gamma \vdash \mathscr{R}T$ or $\Gamma \vdash Tx \veebar \neg Ex$, $\Gamma \vdash \mathscr{R}U$ and $\Gamma \vdash \mathscr{C}V$. According to the inductive hypothesis, given the truth of all members of Γ in $\mathfrak{M}, v$ at w, we obtain the truth of $xTU \sqsupset_x xTV$, $\mathscr{R}T$ if $[\![T]\!]_v^{\mathfrak{F},w'} \neq [\![E]\!]_v^{\mathfrak{F},w'}$ for some w' with wRw', $\mathscr{R}U$ and $\mathscr{C}V$, and hence all the four conditions of Definition 15, case 3d. It is immediate (from the same case of Definition 15) that $[\![V]\!]_v^{\mathfrak{M}} \in [\![\mathscr{R}]\!]_v^{\mathfrak{M},w}$, that is, $\mathscr{R}V$ is true in $\mathfrak{M}, v$ at w. □

To prove the *completeness* of CO, we use some essential features of the completeness proofs in [8, 13, 7] (see also [23]).

The vocabulary of $\mathscr{L}$CO is extended by an infinite number of first-order variables (parameters) $u_1, u_2, u_3, \ldots$ (set $\mathcal{V}^{1'}$) and an infinite number of new second-order constants $F_1, F_2, F_3, \ldots$ (set $\mathcal{A}'$), thus obtaining the language $\mathscr{L}$CO'. Since parameters are not bound by any operator, we say that a formula ϕ of $\mathscr{L}\mathsf{CO}'$ is closed iff no free variable of $\mathcal{V}^1$ or $\mathcal{V}^2$ occurs in ϕ. We say that a set w of closed sentences is saturated iff w is maximal consistent,[34] and ω-complete with respect to the instantiations of a, $\neg a$, $\neg \Box$, and $\neg \vee$ sentences. For example, for each sentence $ax\, SP \in w$, there is a variable u such that $Su \in w$, where u is a new instantiation term at a stage k of the building of worlds w and their sequence W. We add that there should be a denumerable set of new parameters and a denumerable set of new second-order constants for each w.

Let us define, for each consistent set Δ of sentences of $\mathscr{L}\mathsf{CO}$, a set Δ^c, obtained by an injective *mapping c* of first-order variables in $\mathsf{free}(\Delta)$ to $\mathcal{V}^{1'}$ and of second-order variables in $\mathsf{free}(\Delta)$ to $\mathcal{A}'$, still leaving infinitely many new variables in $\mathcal{V}^{1'}$ and new constants in $\mathcal{A}'$. If Δ is consistent, then the set Δ^c is consistent, since the same shape of the proof as for the inconsistency of Δ^c is applicable to prove the inconsistency of Δ (with the replacement of terms corresponding to the substitution c and with a mapping of new instantiating terms in the derivation of the inconsistency of Δ^c to the terms that are new within the derivation of the inconsistency of Δ). We now assume that it could be proved that Δ^c is a subset of a saturated set w of sentences of $\mathscr{L}\mathsf{CO}'$, where w is a member of a $\neg\Box$-, $\neg\sqsupset$-, and $\neg\vee$-complete sequence W of saturated sets (proof mainly along the lines of a Gallinian construction and argument [8]).

Definition 17 (Canonical frame and variable assignment). *Canonical frame $\mathfrak{F}^c = \langle W, R^c, D^c, D(1)^c, q^c, I^c \rangle$, where (a) W is a set of saturated sets (w) of closed sentences of $\mathscr{L}\mathsf{CO}'$, (b) wRw' iff $\sqrt{w} \subseteq w'$, (c) $D^c = \mathcal{V}^{1'}$, (d) $\varnothing \neq D(1)^c \subseteq \wp(D^c)^W$, (e) $q^c \in D(1)$ with $q^c(w) = \{u \in D^c \mid Eu \in w\}$,[35] (f) $I^c(K \in \mathcal{A} \cup \mathcal{A}') \in D(1)^c$, $I^c(E) = q^c$, $I^c(\mathscr{C}, w) \subseteq D(1)^c \setminus \{q^c\}$, and $I^c(\mathscr{R}, w) \subseteq I^c(\mathscr{C}, w)$. For a canonical variable assignment v^c, $v^c(x \in \mathcal{V}^1) \in D^c$, $v^c(u \in \mathcal{V}^{1'}) = u$, and $v^c(X \in \mathcal{V}^2) \in D(1)^c$.*

Definition 18 (Canonical model and variable assignment). *Canonical model $\mathfrak{M}^c$ is a canonical frame satisfying the conditions as in Definition 15 (model) for λ-abstracts, $\mathscr{C}$ and $\mathscr{R}$, and where, in addition, $D(1)^c$ is the range of the interpretation I^c of $\mathcal{A} \cup \mathcal{A}' \cup \{E\}$, and for each constant $K \in \mathcal{A} \cup \mathcal{A}'$, $I^c(K, w) = \{u \in D^c \mid Ku \in w\}$. Canonical variable assignment is defined in the same way as for the canonical frame.*

Note that, due to the ω-completeness, each $\lambda x.\phi$ receives its place and denotation in a canonical model by means of some second-order constant that, for example, instantiates the property of $(\lambda X.Xx \vee_x \neg\phi x)$ occurring in the subject place of a second-order subject-predicate sentence.

---

[34] We extend the concepts of inconsistency and consistency of Definition 9 to the language $\mathscr{L}\mathsf{CO}'$.
[35] We note that $q^c(w)$ is non-empty because of Axiom E and the maximality of w.

We extend the concepts of satisfaction and consequence of definitions 14 and 16 to the language $\mathscr{L}\mathsf{CO}'$.

Let $\phi^c = \phi(u/x, T \in \mathcal{A} \cup \mathcal{A}' \cup \{E\}/X)$, where $u = v^c(x)$, $[\![T]\!]_{v^c}^{\mathfrak{M}^c,w} = v^c(X)$, and ϕ is any expression of $\mathscr{L}\mathsf{CO}$. We call the substitution by means of which ϕ^c is defined *canonical substitution*. Note that ϕ^c is a closed formula possibly containing free first-order variables from $\mathcal{V}^{1'}$ (parameters). Term T^c is defined analogously, by means of the canonical substitution applied to T.

Theorem 4 (Canonical satisfaction in a model). *For each ϕ of $\mathscr{L}\mathsf{CO}$,*

$$\mathfrak{M}^c, w \models_{v^c} \phi \text{ iff } \phi^c \in w.$$

Proof. The first-order basic case (Kx and Ku for $v^c(x) = u$) is obvious on the ground of the definition of canonical model and variable assignment. The second-order basic case ($\mathscr{C}T, \mathscr{R}T$) should be proved on the ground of the special conditions in definitions 18 and 15. We give some examples for complex formulas and formulas of the shape $\mathscr{R}T$.

(a) ($ax SP$, $aX\mathscr{S}\mathscr{P}$). (i) Assumption $(ax SP)^c \in w$ implies that for some $u \in \mathcal{V}^{1'}$, $S^c u \in w$ (ω-completeness of w), and that for each $u' \in \mathcal{V}^{1'}$, if $S^c u' \in w$ then $P^c u' \in w$ (aE and substitutability of u' since u' is a parameter, maximality of w). By the inductive hypothesis, according to definitions 18 and 14, we obtain that $\varnothing \neq [\![S]\!]_{v^c}^{\mathfrak{M}^c,w} \subseteq [\![P]\!]_{v^c}^{\mathfrak{M}^c,w}$, that is, $\mathfrak{M}^c, w \models_{v^c} ax SP$. (ii) Assumption $(ax SP)^c \notin w$ implies that either for each $u \in \mathcal{V}^{1'}$, $S^c u \notin w$ (substitutability of u, maximality of w), or for some $u \in \mathcal{V}^{1'}$, $S^c u \in w$ but $P^c u \notin w$ (by ω-completeness of w). According to the inductive hypothesis, in the canonical model, this amounts to saying that $\mathfrak{M}^c, w \not\models_{v^c} ax SP$. – With some differences, the proof for $aX\mathscr{S}\mathscr{P}$ is analogous with that for $ax SP$. We prove the right to left direction (i). Assume that $aX\mathscr{S}\mathscr{P}^c \in w$. It follows that for some $K \in \mathcal{A} \cup \mathcal{A}' \cup \{E\}$, $\mathscr{S}^c K \in w$ (ω-completeness of w). Also, for each $K' \in \mathcal{A} \cup \mathcal{A}' \cup \{E\}$, if $\mathscr{S}^c K' \in w$ then $\mathscr{P}^c K' \in w$ (aE and substitutability of K', maximality of w). By the inductive hypothesis, and since the set $\mathcal{A} \cup \mathcal{A}' \cup \{E\}$, by the meanings of its members, "covers" the second-order domain $D(1)^c$ (see Definition 18), we conclude that $\mathfrak{M}^c, w \models_{v^c} aX\mathscr{S}\mathscr{P}$ (Definition 14).

(b) ($\phi \sqsupset_{(\alpha)} \psi$). (i) From the assumption $(\phi \sqsupset_{(\alpha)} \psi)^c \in w$ it follows: for each w' such that $\sqrt{w} \subseteq w'$, if $\phi^c(\gamma/\alpha) \in w'$ then $\psi^c(\gamma/\alpha) \in w'$, for any $\gamma \in \mathcal{V}^{1'}$ ($\gamma \in \mathcal{A} \cup \mathcal{A}' \cup \{E\}$) substitutable in $(\phi \sqsupset \psi)^c$ (MP, maximality of each w'). In accordance with the inductive hypothesis and definitions 18 and 14, it follows that $\mathfrak{M}^c, w \models_{v^c} \phi \sqsupset_{(\alpha)} \psi$. (ii) Assume that $(\phi \sqsupset_{(\alpha)} \psi)^c \notin w$. Then, for some $\gamma \in \mathcal{V}^{1'}$ ($\gamma \in \mathcal{A} \cup \mathcal{A}' \cup \{E\}$) (substitutability of γ in $(\phi \sqsupset \psi)^c$) and some w'

such that $\sqrt{w} \subseteq w'$, $\phi^c(\gamma) \in w'$ but $\psi^c(\gamma) \notin w'$, which, under the inductive hypothesis and in the canonical domain gives $\mathfrak{M}^c, w \not\models_{v^c} \phi \sqsupset_{(\alpha)} \psi$.

(c) $((\lambda\alpha.\phi)\gamma)$. Obvious from definitions 14 (case 1) and 15 (case 1), the inductive hypothesis on $\phi(\gamma)$, and the maximality of w.

(d) $(\mathscr{R}T)$. (i) Let us first consider the intersection case (see Definition 15, case 3b, and Definition 18). Note that $[\![T]\!]_{v^c}^{\mathfrak{F}^c,w} = \bigcap\{[\![U]\!]_{v^c}^{\mathfrak{F}^c,w} \mid [\![U]\!]_{v^c}^{\mathfrak{F}^c} \in [\![\mathscr{R}]\!]_{v^c}^{\mathfrak{F}^c,w}\}$ defines T as $\lambda x.a X \mathscr{R}(\lambda X.Xx)$, that is, as R. According to definitions 15 and 18, $\mathfrak{M}^c, w \models_{v^c} \mathscr{R}T$, and hence $\mathfrak{M}^c, w \models_{v^c} \mathscr{R}R$. But $\mathscr{R}R \in w$ since $\mathscr{R}R$ is an axiom (maximality of w). Also, $R^c = R$ (Definition 7). Therefore, $\mathfrak{M}^c, w \models_{v^c} \mathscr{R}R$ iff $\mathscr{R}R^c \in w$. (The same holds for any T semantically and syntactically equivalent to R). In addition, we remark that the clause $[\![T]\!]_{v}^{\mathfrak{F}^c,w'} \subseteq [\![T]\!]_{v^c}^{\mathfrak{F}^c,w''}$ (with $w'Rw''$) accounts for Axiom A4. (ii) Let us take the consequent case (see Definition 15, case 3d, and Definition 18) as another example. The semantic antecedent conditions can be expressed as $\mathfrak{M}^c, w \models_{v^c} xTU \sqsupset_x xTV$, $\mathfrak{M}^c, w \models_{v^c} \mathscr{R}T$ if $[\![T]\!]_{v}^{\mathfrak{F},w'} \neq [\![E]\!]_{v}^{\mathfrak{F},w'}$ for some w' with wRw', $\mathfrak{M}^c, w \models_{v^c} \mathscr{R}U$, $\mathfrak{M}^c, w \models_{v^c} \mathscr{C}V$ and their consequent as $\mathfrak{M}^c, w \models_{v^c} \mathscr{R}V$. Let us correspondingly suppose that $(xTU \sqsupset_x xTV)^c, \mathscr{R}U^c, \mathscr{C}V^c \in w$, and $\mathscr{C}T^c \in w$ if $Tx \veebar_x \neg Ex \notin w$. Then, on the ground of RC, $\mathscr{R}V^c$ is implied by w and thus (by maximality) $\mathscr{R}V^c \in w$. Therefore, on the above semantic and membership conditions, $\mathfrak{M}^c, w \models_{v^c} \mathscr{R}V$ iff $\mathscr{R}V^c \in w$. $\square$

Proposition 10. *Let $\mathfrak{M}_\Delta^c$ be a canonical model with the sequence W built on the consistent set $\Delta^c \subseteq w$ by some substitution c (see above on the construction of Δ^c) applied to a consistent set Δ of sentences of $\mathcal{L}$CO. Let v_Δ^c be a canonical variable assignment such that $v_\Delta^c(x \in \mathsf{free}(\Delta)) = c(x)$ and $v_\Delta^c(X \in \mathsf{free}(\Delta)) = I_\Delta^c(c(X))$. Accordingly,*

$$\mathfrak{M}_\Delta^c, w \models_{v_\Delta^c} \Delta.$$

Proof. The proposition follows from Theorem 4 and the fact that $\Delta^c \subseteq w$. $\square$

Theorem 5 (Completeness). *If $\Gamma \models \phi$ then $\Gamma \vdash \phi$, where ϕ and each $\psi \in \Gamma$ are formulas of $\mathcal{L}$CO.*

Proof. Let Δ be a consistent set of formulas of $\mathcal{L}$CO. As already mentioned, if so, then the set Δ^c is consistent. Since $\Delta^c \subseteq w$ for some w in W on the ground of which the canonical model $\mathfrak{M}_\Delta^c$ and variable assignment v_Δ^c can be defined, Δ is satisfied by $\mathfrak{M}_\Delta^c$ and v_Δ^c (Proposition 10). Thus, there are a model $\mathfrak{M}$ and a variable assignment v, restricting $\mathfrak{M}_\Delta^c$ and v_Δ^c to the interpretation and variable assignment of the symbols of $\mathcal{L}$CO, such that $\mathfrak{M}$ and v satisfy Δ. Therefore, by contraposition, from the unsatisfiability of $\Gamma \cup \{\neg\phi\}$ by any model $\mathfrak{M}$ and variable assignment v the inconsistency of $\Gamma \cup \{\neg\phi\}$ follows, which establishes the theorem. $\square$

A concluding remark

It is often stated that Gödel's ontological proof of the existence of God significantly relies on Leibniz's ontological proof, especially by including, in its first part, an argument for the possibility of the existence of God. Although this view is certainly justified in many important aspects, we suggest that it needs to be complemented by an elaboration of the possible interconnections of Gödel's proof with Kant's criticism of ontotheology. The more so since we know that Kant's philosophy was a rather constant reference point for Gödel, be it in a critical or an inspiring way, from the earliest time of his intellectual development (see [35, pp. 68–69], [11, 19]). The account and formalization described above are thought as a proposal to possibly bridge the seeming philosophical as well as technical gap between Kant's and Gödel's approaches. Although this may imply some difficulties or peculiarities in the formal language and system (which are meant to preserve Kantian logical forms and to satisfy the requirements of modern proof rigour), such an approach might, hopefully, broaden our insights about both philosophers and about ontotheological concepts involved.

References

[1] Theodora Achourioti and Michiel van Lambalgen. A formalization of Kant's transcendental logic. *Review of Symbolic Logic*, 4(2):254–289, 2011.

[2] C. Anthony Anderson. Some emendations of Gödel's ontological proof. *Faith and Philosophy*, 7(3):291–303, 1990.

[3] Johannes Czermak. Abriß des ontologischen Argumentes. In Eckehart Köhler, Bernd Buldt, et. al., editors, *Kurt Gödel: Wahrheit und Beweisbarkeit*, vol. 2, pages 309–324. Wien: Öbv et Hpt, 2002.

[4] Johannes Czermak. From ontological proofs to theological theories. In Kordula Świętorzecka, editor, *Gödel's ontological argument: history, modifications, and controversies*, pages 47–83. Warszawa: Semper, 2015.

[5] René Descartes. *Meditationes de prima philosophia*. Charles Adam and Paul Tannery, editors, *Œuvres de Descartes*, vol. 7. Paris: Cerf, 1904.

[6] Melvin Fitting. *Types, Tableaus, and Gödel's God*. Dordrecht, etc.: Kluwer, 2002.

[7] Melvin Fitting. Possible world semantics for first-order LP. CUNY Academic Works, TR2011010. 2011. – Revised: Possible world semantics for first-order logic of proofs, *Annals of Pure and Applied Logic*, 165:225–240, 2014.

[8] Daniel Gallin. *Intensional and Higher-Order Modal Logic: With Applications to Montague Semantics*. Amsterdam, Oxford, etc.: North-Holland, Elsevier, 1975.

[9] Kurt Gödel. *Collected Works*. Solomon Feferman et al., editors, vol. 1–5. Oxford University Press, 1986–2003.

[10] Kurt Gödel. What is Cantor's continuum problem? In [9], vol. 2, pages 254–270. 1990.

[11] Kurt Gödel. The modern development of the foundations of mathematics in the light of philosophy. In [9], vol. 3, pages 374–387. 1995.

[12] Kurt Gödel. Ontological proof / Appendix B: Texts relating to the ontological proof. In [9], vol. 3, pages 403–404, 429–437. 1995.

[13] Petr Hájek. Magari and others on Gödel's ontological proof. In Aldo Ursini and Paolo Agliano, editors, *Logic and Algebra*, pages 125–135. New York, etc.: Dekker, 1996.

[14] Petr Hájek. Gödel's ontological proof and its variants. In Matthias Baaz and others, editors, *Kurt Gödel and the Foundations of Mathematics: Horizons of Truth*, pages 307–321. Cambridge University Press, 2011.

[15] Immanuel Kant. *Logik*. Königlich Preußische Akademie der Wissenschaften, editors, *Kant's Gesammelte Schriften*, vol. 16. Berlin, Leipzig: de Gruyter, 1924.

[16] Immanuel Kant. *Kritik der reinen Vernunft, 2. Aufl. Kants Werke*, vol. 3. Berlin: de Gruyter, 1968. (1st edition in vol. 4). – Transl. by Paul Guyer and Allen W. Wood, *Critique of Pure Reason*, Cambridge University Press, 1998.

[17] Immanuel Kant. *Prolegomena zu einer jeden künftigen Metaphysik. Kants Werke*, vol. 4. Berlin: de Gruyter, 1968. – Transl. and ed. by Gary Hatfield, *Prolegomena to Any Future Metaphysics*, Cambridge University Press, 2004.

[18] Immanuel Kant. *Lectures on Logic*. Transl. and ed. by J. Michael Young. Cambridge University Press, 1992.

[19] Srećko Kovač. Gödel, Kant, and the path of a science. *Inquiry*, 51(2):147–169, 2008.

[20] Srećko Kovač. In what sense is Kantian principle of contradiction non-classical? *Logic and Logical Philosophy*, 17(3):251–274, 2008.

[21] Srećko Kovač. Modal collapse in Gödel's ontological proof. In Mirosław Szatkowski, editor, *Ontological Proofs Today*, pages 323–343. Frankfurt, etc.: Ontos, 2012.

[22] Srećko Kovač. Forms of judgment as a link between mind and the concepts of substance and cause. In Mirosław Szatkowski and Marek Rosiak, editors, *Substantiality and Causality*, pages 51–66. Boston, Berlin, etc.: de Gruyter, 2014.

[23] Srećko Kovač. Causal interpretation of Gödel's ontological proof. In Kordula Świętorzecka, editor, *Gödel's ontological argument*, pages 163–201. See [4]. 2015.

[24] Srećko Kovač. Concepts, space-and-time, metaphysics. In Mirosław Szatkowski, editor, *God, Time, Infinity*, pages 61–85. Berlin, Boston: de Gruyter, 2018.

[25] Béatrice Longuenesse. *Kant and the Capacity to Judge*. Transl. by Charles T. Wolfe. Princeton University Press, 1998.

[26] Gottfried Wilhelm Leibniz. *Quod ens perfectissimum existit*. In C. I. Gerhardt, editor, *Die philosophischen Schriften*, vol. 7, pages 261–262. Hildesheim, New York: Olms, 1978.

[27] Marek Nasieniewski. *Logiki zdaniowe wyrażalne przez modalność*. Toruń: Wydawnictwo Naukowe Uniwersytetu Mikołaja Kopernika, 2011.

[28] Jerzy Perzanowski. Ontological arguments II: Cartesian and Leibnizian. In Hans Burkhardt and Barry Smith, editors, *Handbook of Metaphysics and Ontology*, vol. 2, pages 625–633. München: Philosophia Verlag, 1991.

[29] Ian Proops. Kant on the ontological argument. *Noûs*, 49(1):1–27, 2015.

[30] Klaus Reich. *Die Vollständigkeit der kantischen Urteilstafel*, 2nd edition. Berlin: Schoetz, 1948.

[31] Jordan Howard Sobel. *Logic and Theism*. Cambridge University Press, 2004. – A revised version from 2006 was available at `http://www.scar.utoronto.ca/~sobel/Logic_Theism/`.

[32] Kordula Świętorzecka. O pewnych formalnych założeniach semantycznych niektórych sformalizowanych argumentów ontologicznych. *Studia philosophiae christianae*, 38(2):55–86, 2002.

[33] Kordula Świętorzecka and Marcin Łyczak. An even more Leibnizian version of Gödel's ontological argument. *Journal of Applied Logics*, this issue.

[34] Mary Tiles. Kant: from general to transcendental logic. In Dov M. Gabbay and John Woods, editors, *Handbook of the History of Logic*, vol. 3, pages 85-130. Amsterdam, etc.: Elsevier, 2004.

[35] Hao Wang. *A Logical Journey: From Gödel to Philosophy*. Cambridge [Mass.]: The MIT Press, 1996.

An even more Leibnizian version of Gödel's ontological argument

Kordula Świętorzecka and Marcin Łyczak
Cardinal Stefan Wyszyński University in Warsaw, Poland
Institute of Philosophy
k.swietorzecka@uksw.edu.pl
m.lyczak@uksw.edu.pl

Abstract

We propose a modification of Gödel's ontological argument for God's existence from his 'Ontologischer Beweis' manuscript (1970). We follow a Leibnizian onto-theology, especially two of Leibniz's letters from 1676 and 1677, to which Gödel could relate. We consider two differences between Gödel and Leibniz. We argue for the superiority of Leibniz's ideas, while preserving the main structure of the Gödelian argument. Our first aim is to bring Gödel's concept of positiveness closer to the idea of a Leibnizian *perfectio* which should not be understood via *negations*. Our second goal is to analyze the concept of being necessary in terms of a Leibnizian *demonstrability*. To this end, we formulate an S4 version of Gödel's argument without using negative predicate terms. We sketch a model for our theory that allows us to express a few specific properties of the Leibnizian God.

Keywords: ontological argument, existence of God, Gödel, Leibniz, logical philosophy, logic applied to ontology

Introduction

We propose a modification of Gödel's argument for God's existence, originally expressed in his 1970 manuscript 'Ontologischer Beweis' ([7], OB). Our proposal follows a Leibnizian onto-theology, which Gödel referred to in many of his philosophical writings.

We claim that Gödel's approach combines the main ideas of two Leibnizian writings from 1676 and 1677, respectively. The first one, titled 'That the most perfect being exists', was included in a letter to Spinoza ([12, 427], trans. in [13, 167-169]), and is usually considered by commentators of OB as the main source for Gödel's view. The second source comes from Leibniz's correspondence with Eckhard ([12, 588] trans. in [1, 136-137]), in which Leibniz addresses the question of God's existence in connection with the concept of God's essence. This feature is also present and significant in Gödel's argument. It is well known, that Leibniz's letter to Spinoza influenced Gödel's formulation of the proof of the Cartesian lemma, which is one of two main lemmata of the key thesis of God's necessary existence. It states that the existence of the most perfect being is possible. The second pillar of Gödel's ontological argument is the Leibnizian lemma according to which the possible existence of God implies His necessary existence. The proof of this second lemma is based on the mentioned concept of essence, which is understood quite differently from Leibniz's interpretation.

A comparative analysis of OB and Leibniz's texts has already been carried out in [2] and [16]. Here we want to consider two differences between Gödel's and Leibniz's approaches. We will argue for the superiority of Leibniz's approach, while preserving the main structure of the Gödelian argument.

Our first aim is to bring the key concept of *positiveness* used by Gödel closer to Leibniz's idea of *perfectio*, which underlies the ontological argument from 1676. It is true that Leibniz considered perfections as *simple*, *positive*, and *absolute qualities*, whereas Gödel's concept of positiveness comprises just *properties* (including complex and relational ones, which do not qualify as Leibnizian perfections). However, our aim is to impose on Gödel's positive properties at least the condition accepted by Leibniz for positive qualities. The latter are not to be "understood through negations" ([13, 167-169]). Following this restriction, we modify Gödel's proof in such a way as to eliminate negative predicate terms from it. The language of the whole theory still includes term negation, but in the sense of contrariness.

Our second modification is also inspired by Leibniz's letter to Spinoza. This time we want to consider the original idea of a connection between the concept of *necessity* and the concept of *demonstrability*. Leibniz explicitly understands necessarily true propositions that are not known per se as demonstrable ones. This motivates us to use a formal basis for OB different from the one that is usually considered. It is generally assumed that Gödel's approach falls within the framework of logic S5.[1] However, the concept of demonstrability does not fulfill axiom *5*; rather, it fulfills axiom *4*. By combining Leibnizian modalities in their proof-theoretical meaning with

---

[1] A survey of the logics used in various formalizations of OB is given in [16, 25-29]. Modal frames were investigated earlier by S. Kovač in [10]. In addition to S5, the mostly used modal logics which do not require many changes in the original structure of the argument in OB are K5, KD45, and KB.

S4 modalities we follow a proposal advanced by Adams [1, 46-50].[2] Our approach further employs the idea of grounding OB on the S4 system introduced in [15]. That formalism is a 'compromise' between S4 and S5 modalities. Our formal frame is logic S4, but a specific instantiation of 5 is needed as an axiom to derive the Leibnizian lemma. As a result we do not claim that S4 exclusively describes the modalities of Leibniz's texts. It may be worth mentioning that our proposal also addresses Gödel's doubts about "using some principle in modal logic" in OB as being too strong – Adams supposes that this principle is the axiom 5 [2, 391].

We formulate a new version of OB, which also offers interesting references to some of Leibniz's views on the 'nature' of God. It turns out that our theory has models in which God's nature is not maximal, in the sense that there are properties such that God does not possess them or their contraries (negations). This corresponds with the Leibnizian concept of God, Who is neither finite nor infinite in time and space. Secondly, in our proposal God is unchanging in the sense that all of His properties are His attributes, but He is not determined by at least some relative properties, that He only possibly possesses (like "being the creator of any possible but not actual world").

We begin our discussion with D. Scott's presentation of Gödel's argument with an explicit description of the assumed formal system (1). Next, we show the possibility of eliminating negative terms from the ontological proof and highlight certain derivative dependencies between this proposal and certain axioms of Scott's theory (2). Finally, we introduce the axiomatic description of our modification of OB, sketch its semantics and show a few interesting semantical observations within this new framework (3).

1 Starting point: the formalization of OB by D. Scott

The original OB manuscript was discussed by Gödel with Scott, who presented it during his seminars in 1970. As a result of these exchanges we now have the so-called Scott version of Gödel's argument, which is considered the version closest to the original ideas of the manuscript. We accept the Scott version as an adequate supplement to Gödel's approach.

For the convenience of the reader we provide a short presentation of the Scott theory, following its description in [16, 17-24].

---

[2]Interestingly, the formula (4) $\Box A \to \Box\Box A$ follows from the law about the *rationality of necessary statements*, which is also attributed to Leibniz by Perzanowski. If we assume that contingency is described by $\overline{K}A =: \Diamond A \land \Diamond \neg A$, then we can express the statement "Everything that is necessary, is not contingent" as: $\neg \overline{K} \Box A$. Using the standard definition: $\Diamond A \leftrightarrow \neg\Box\neg A$, we obtain precisely $\Box A \to \Box\Box A$ (cf. [14, 99]).

The vocabulary of the used symbolic language consists of: individual variables $x, y, z, \ldots$; first-order unary predicate variables $\varphi, \psi, \chi, \ldots$; constants G (*is God*), and NE (*is necessarily existent*); first order identity predicate $=$; second-order predicate P (*is positive*); symbols: $^{-}$ (for term negation), $\neg, \rightarrow, \forall, \Box$; and parentheses.

We accept the following definitions of predicate terms and formulas:

$$\tau ::= \varphi \mid \overline{\varphi} \mid G \mid NE$$

$$A ::= \tau x \mid x = x \mid \mathsf{P}(\tau) \mid \neg A \mid A \rightarrow A \mid \forall x A \mid \forall \varphi A \mid \Box A$$

Symbols $\wedge, \vee, \leftrightarrow, \exists$ are defined in the standard way.

We use α, β as representing individual or predicate variables.

The Scott theory is based on a logic characterized by:

- all classical sentential tautologies (PC)
- formulas of the following shapes:

$(Q2)$ $\forall \alpha A \rightarrow A(^{\tau}/_{\alpha})$ τ is substitutable for α.

$(id1)$ $x = x$

$(id2)$ $A \wedge x = y \rightarrow A[^{y}/_{x}]$

(K) $\Box(A \rightarrow B) \rightarrow (\Box A \rightarrow \Box B)$

(T) $\Box A \rightarrow A$

(5) $\Diamond \Box A \rightarrow \Box A$

$(\Diamond/\Box)$ $\Diamond A \leftrightarrow \neg \Box \neg A$

The primitive rules are modus ponens; rules for introducing quantifiers: $\vdash A \rightarrow B \implies \vdash A \rightarrow \forall \alpha B$, where α is not free in A; and the necessitation rule: $\vdash A \implies \vdash \Box A$ $(R\Box)$.

The above system is called Q2S5.

We add to Q2S5 $\Box$ closures of the following equivalences:

$(^{-})$ $\overline{\tau}x \leftrightarrow \neg \tau x$ \hfill (*negative property*)

$(\boldsymbol{G})$ $Gx \leftrightarrow \forall \varphi(\mathsf{P}(\varphi) \rightarrow \varphi x)$ \hfill (*God*)

$(\boldsymbol{NE})$ $NEx \leftrightarrow \forall \varphi(\varphi Ess.x \rightarrow \Box \exists y \varphi y)$ \hfill (*necessary existence*)

where $\varphi Ess.x$ means $\varphi x \wedge \forall \psi(\psi(x) \rightarrow \Box \forall y(\varphi(y) \rightarrow \psi(y)))$ \hfill (***Ess.***)

The expression $\varphi Ess.x$ reads as follows: *property φ is an essence of x*.
In Q2S5 extended by a modal version of the comprehension schema:

(MCS) $\exists \varphi \Box \forall x (\varphi x \leftrightarrow A(x))$, φ is not in A

our $\Box$ closures of $^-$, $\boldsymbol{G}$, $\boldsymbol{NE}$ are cases of MCS.

The following are specific axioms for P:

$(A1)$ $\mathsf{P}(\varphi) \leftrightarrow \neg \mathsf{P}(\overline{\varphi})$

$(A2)$ $\mathsf{P}(\varphi) \wedge \varphi \subset_\Box \psi \to \mathsf{P}(\psi)$, where $\varphi \subset_\Box \psi$ means $\Box \forall x(\varphi x \to \psi x)$

$(A3)$ $\mathsf{P}(G)$

$(A4)$ $\mathsf{P}(\varphi) \to \Box \mathsf{P}(\varphi)$

$(A5)$ $\mathsf{P}(NE)$

The Scott theory is just the extension of Q2S5 by the set of axioms $A1$-$A5$ above and the set of $\Box$ closures of the mentioned cases of the comprehension schema $\Box(^-$, $\boldsymbol{G}$, $\boldsymbol{NE})$. We call it $\mathsf{GO}^\circ = \mathsf{Q2S5}[A1\text{-}A5, \Box(^-, \boldsymbol{G}, \boldsymbol{NE})]$.

Gödel's ontological argument may now be reconstructed in a simple way.

We start with the theorem about the possible existence of the subject of any positive properties:

Th 1. $\mathsf{P}(\varphi) \to \Diamond \exists x \varphi x$

Proof. 1. $\mathsf{P}(\varphi) \wedge \varphi \subset_\Box \overline{\varphi} \to \mathsf{P}(\overline{\varphi})$ [$A2$], 2. $\mathsf{P}(\varphi) \wedge \Box \forall x(\varphi x \to \overline{\varphi} x) \to \neg \mathsf{P}(\varphi)$ [1, $A1$, $\subset_\Box$], 3. $\mathsf{P}(\varphi) \wedge \Box \forall x(\varphi x \to \neg \varphi x) \to \neg \mathsf{P}(\varphi)$ [2,$^-$], 4. $\mathsf{P}(\varphi) \to \neg \Box \forall x(\varphi x \to \neg \varphi x)$ [3, PC], 5. $\mathsf{P}(\varphi) \to \Diamond \exists x \varphi x$ [4, $\Diamond/\Box$]

Next, we derive the counterpart of the Cartesian lemma:

$(\mathbf{CL})$ $\Diamond \exists x Gx$ [Th1, $A3$]

The Leibnizian lemma is obtained through the thesis that the property of being God is essential to Him:

Th 2. $Gx \to GEss.x$

Proof. 1. $\Box \mathsf{P}(\varphi) \to \Box \forall x(Gx \to \varphi x)$ [$\Box \boldsymbol{G}$, K], 2. $\mathsf{P}(\varphi) \to \Box \forall x(Gx \to \varphi x)$ [1, A4], 3. $Gx \to (\varphi x \to \mathsf{P}(\varphi))$ [$\boldsymbol{G}$, $A1$,$^-$], 4. $Gx \to (\varphi x \to \Box \forall x(Gx \to \varphi x))$ [PC, 2, 3], 5. $Gx \to Gx \wedge \forall \varphi (\varphi x \to \Box \forall x(Gx \to \varphi x))$ [Q2,4]. 6. $Gx \to GEss.x$ [**Ess.**, 5]

The Leibnizian lemma is expressed as follows:

$(\mathbf{LL})$ $\Diamond \exists x Gx \to \Box \exists x Gx$

Proof. 1. $Gx \to NEx$ [$\boldsymbol{G}$, $A5$], 2. $Gx \to \forall\varphi(\varphi Ess.x \to \Box\exists x\varphi x)$ [$\boldsymbol{NE}$, 1], 3. $Gx \to (GEss.x \to \Box\exists xGx)$ [2, $Q2$], 4. $Gx \to \Box\exists xGx$ [Th2, 3], 5. $\exists xGx \to \Box\exists xGx$ [$Q2$, 4], 6. $\Diamond\exists xGx \to \Diamond\Box\exists xGx$ [$\vdash A \to B \Longrightarrow \vdash \Diamond A \to \Diamond B$, 5]³, 7. $\Diamond\exists xGx \to \Box\exists xGx$ [5, 6]

Finally we derive the key thesis from **CL** and **LL**

    Th 3. $\Box\exists xGx$

Theory GO° has further interesting theses:

Th 4.	$Gx \to \forall\varphi(\varphi x \to \mathsf{P}(\varphi))$	($\boldsymbol{G}$, ⁻, $A4$)
Th 5.	$\Diamond\mathsf{P}(\varphi) \to \mathsf{P}(\varphi)$	($A4$, $A1$)
Th 6.	$\varphi Ess.x \land \psi Ess.x \to \Box\varphi = \psi$	(**Ess.**)
	where $\varphi = \psi$ stands for $\forall x(\varphi x \leftrightarrow \psi x)$	
Th 7.	$\varphi Ess.x \land \varphi Ess.y \to \forall\psi(\psi x \leftrightarrow \psi y)$	(**Ess.**, T)
Th 8.	$Gx \land Gy \to \forall\psi(\psi x \leftrightarrow \psi y)$	(Th2, Th7)

We can also add to GO° all $\Box$ closures of

    ($\boldsymbol{I_x}$) $I_x y \leftrightarrow x = y$ (*identity in relation to x*)

In GO°[$\Box\boldsymbol{I_x}$] we can next prove the following further theses:

Th 9.	$\varphi Ess.x \land \varphi Ess.y \to x = y$	(Th7, $\boldsymbol{I_x}$, $id1$)
Th 10.	$Gx \land Gy \to x = y$	(Th8, $\boldsymbol{I_x}$, $id1$)
Th 11.	$Gx \to \Box\forall y(Gy \to x = y)$	(**Ess.**, $\boldsymbol{I_x}$, $id1$, Th2)
Th 12.	$Gx \to \Box Gx$	(Th11, $id1$, Th3)
Th 13.	$\exists x\Box Gx$	(Th12, Th3, T)
Th 14.	$Gx \to \forall\varphi(\varphi x \to \Box\varphi x)$	($\Box\boldsymbol{G}$, K, $A4$, Th12, Th4)
Th 15.	$Gx \to \forall\varphi(\Diamond\varphi x \to \varphi x)$	(Th14, ⁻)

We will refer to some of these theses in our further considerations.

    A description of an adequate semantics for GO°[$\Box\boldsymbol{I_x}$] is given in [16, 23-24] (which follows [9]). We will adopt this semantics when formulating our proposal in section 3.

---

³The rule $\vdash A \to B \Longrightarrow \vdash \Diamond A \to \Diamond B$ is derivable from $R\Box$. The theory GO° is $\Box$ transparent in the sense that for every A which is GO° thesis we get $A \to \Box A$ only with modus ponens and rules for $\forall$ [5, 317].

2 Negative terms are not needed in the argument

A significant difference between Leibnizian perfections and Gödel's positive properties, which we want to eliminate in favor of the former, concerns the use of negative predicate terms. These are introduced in GO° with the equivalence $^-$, which also enables us to describe positive properties as complements (negations) of others. Leibnizian perfections, instead, are not describable via negations. As we shall see, the elimination of negative predicate terms does not turn Gödel's argument into a theory weaker than GO°.

Let us consider two more GO° theses:

($\boldsymbol{G^{\leftrightarrow}}$) $Gx \leftrightarrow \forall\varphi(\mathsf{P}(\varphi) \leftrightarrow \varphi x)$ ($\boldsymbol{G}$, Th4)

($\neg\mathsf{P}$) $\exists\varphi\neg\mathsf{P}(\varphi)$ ($A3$, $A1$)

We call $^+$Q2S5 the logic expressed in the language of Q2S5 but with no negative terms and characterized in the same way as Q2S5 without the case $^-$ of the comprehension schema. The labels $^+$Q2$\Diamond$ and $^+$Q2$\Diamond$KT are used for the classical second order fragment of $^+$Q2S5 extended by $\Diamond/\Box$ and $\Diamond/\Box$, K, T, respectively.

It is now to be observed that the formula concerning the possible existence of the subject of any positive properties – Th1 – is derivable from $\neg\mathsf{P}$, and $A2$, which are added to $^+$Q2$\Diamond$:

Fact 2.1 $^+$Q2$\Diamond[\neg\mathsf{P}, A2] \vdash \mathsf{P}(\varphi) \to \Diamond\exists x \varphi x$.

Proof. 1. $\mathsf{P}(\varphi) \land \Box\forall x(\varphi x \to \psi x) \to \mathsf{P}(\psi)$ [$A2, \subset_\Box$], 2. $\mathsf{P}(\varphi) \land \neg\mathsf{P}(\psi) \to \Diamond\exists x(\varphi x \land \neg\psi x)$ [1], 3. $\neg\mathsf{P}(\psi) \to (\mathsf{P}(\varphi) \to \Diamond\exists x \varphi x)$ [2], 4. $\exists\psi\neg\mathsf{P}(\psi) \to (\mathsf{P}(\varphi) \to \Diamond\exists x \varphi x)$ [3], 5. $\mathsf{P}(\varphi) \to \Diamond\exists x \varphi x$ [4, $\neg\mathsf{P}$]

The formula stating that the God-like property is the essence of a subject of this property – Th 2 – is derivable from A4 and $\Box \boldsymbol{G^{\leftrightarrow}}$ in $^+$Q2$\Diamond$KT:

Fact 2.2 $^+$Q2$\Diamond$KT$[A4, \Box\forall \boldsymbol{G^{\leftrightarrow}}] \vdash Gx \to GEss.x$.

Proof. 1. $\Box\mathsf{P}(\varphi) \to \Box\forall x(Gx \to \varphi x)$ [$\Box\forall \boldsymbol{G^{\leftrightarrow}}$, K][4], 2. $\mathsf{P}(\varphi) \to \Box\forall x(Gx \to \varphi x)$ [1, $A4$], 3. $Gx \to (\varphi x \to \mathsf{P}(\varphi))$ [$\Box \boldsymbol{G^{\leftrightarrow}}$, T], 4. $Gx \to (\varphi x \to \Box\forall x(Gx \to \varphi x))$ [PC, 2, 3], 5. $Gx \to Gx \land \forall\varphi(\varphi x \to \Box\forall x(Gx \to \varphi x))$ [$Q2, 4$]. 6. $Gx \to GEss.x$ [**Ess.**, 5].

Following the derivations of **CL** and **LL** in GO° and in view of facts 2.1 and 2.2, we can see that the proof of the key thesis Th3 need not be dependent on negative terms:

---

[4]We need $\Box$ closure of general closure of $\boldsymbol{G^{\leftrightarrow}}$ because we are not in S5.

Fact 2.3 $^+$Q2S5[¬P, $A2 - A5$, $\Box(\boldsymbol{G}^{\leftrightarrow}, \boldsymbol{NE})$] ⊢ $\Box\exists x Gx$.

Of course, a language with no negative terms gives less ability to express many theses contained in GO°. For instance, the formula $A1$ is not expressible in this language. Interestingly, however, the addition of ($^-$) to $^+$Q2S5[¬P, $A2$-$A5$, $\Box(\boldsymbol{G}^{\leftrightarrow}, \boldsymbol{NE})$] changes this situation.

The formula $A1$ is now derivable in Q2S5[¬P, $A2$-$A5$, $\Box(^-, \boldsymbol{G}^{\leftrightarrow}, \boldsymbol{NE})$]:

Fact 2.4 Q2S5[¬P, $A2$-$A5$, $\Box(^-, \boldsymbol{G}^{\leftrightarrow}, \boldsymbol{NE})$] ⊢ $P(\varphi) \leftrightarrow \neg P(\overline{\varphi})$.[5]

Proof. (←) 1. $Gx \to (\varphi x \vee \overline{\varphi}x)$ [$^-$], 2. $Gx \to (P(\overline{\varphi}) \leftrightarrow \overline{\varphi}x)$ [$\boldsymbol{G}^{\leftrightarrow}, \varphi/\overline{\varphi}$], 3. $Gx \to (\neg\varphi x \to \overline{\varphi}x)$ [1], 4. $Gx \to (\neg\varphi x \to P(\overline{\varphi}))$ [3, 2], 5. $Gx \to (\neg P(\overline{\varphi}) \to \varphi x)$ [4], 6. $Gx \to (\neg P(\overline{\varphi}) \to P(\varphi))$ [5, $\boldsymbol{G}^{\leftrightarrow}$], 7. $\exists x Gx \to (\neg P(\overline{\varphi}) \to P(\varphi))$ [5, $\boldsymbol{G}^{\leftrightarrow}$, 6], 8. $\neg P(\overline{\varphi}) \to P(\varphi)$ [5, $\boldsymbol{G}^{\leftrightarrow}$, 7, $\exists x Gx$]

(→) 1. $P(\varphi) \wedge P(\overline{\varphi}) \to (Gx \to (\varphi x \wedge \overline{\varphi}x))$ [$\boldsymbol{G}^{\leftrightarrow}$], 2. $P(\varphi) \wedge P(\overline{\varphi}) \to (Gx \to (\varphi x \wedge \neg\varphi x))$ [2, $^-$], 3. $P(\varphi) \wedge P(\overline{\varphi}) \to \neg Gx$ [2], 4. $P(\varphi) \wedge P(\overline{\varphi}) \to \forall x \neg Gx$ [3], 5. $P(\varphi) \to \neg P(\overline{\varphi})$ [4, $\neg\forall x \neg Gx$]

However, the acceptability of $A1$ is questioned on philosophical grounds and Gödel himself considered the possibility to understand the concept of positiveness without assuming $A1$.[6] Our suggestion is to keep ¬P as an axiom, due to its being more intuitive than $A1$, without impoverishing the language of the formalism by excluding from it the possibility of using negative terms. Following this idea, we accept the weaker meaning of ($^-$) as contrary negation, taking as an axiom the followin:g

($\overset{\to}{^-}$) $\varphi x \to \neg\overline{\varphi}x$

From this, only the following can be proved:

($A1^{\to}$) $P(\varphi) \to \neg P(\overline{\varphi})$ [see 2.4 →]

This is generally considered an acceptable part of $A1$, and it is also assumed by A. Anderson in his simplification of Gödel's argument [3].

A falsification of $A1^{\leftarrow}$ will be shown at the end of the next section. The weakening of ($^-$) to ($\overset{\to}{^-}$) yields another interesting result. Since the formula $\forall\varphi(\varphi x \vee \overline{\varphi}x)$ is not valid in our semantics, we can give models in which, for some properties, God does not possess them nor their negations.

---

[5]The inference from $\neg P(\overline{I})$, where: $Ix \leftrightarrow x = x$, and $A2$ to Th1 was elaborated by J. Czermak [5, 316]. C. Christian has shown that the Scott theory is deductively equivalent to the theory which comes from it by replacing $A1$ with $\neg P(\overline{I})$ [4, 6-7]. Our observations 2.1 and 2.4 use the formula $\exists\varphi\neg P(\varphi)$ which is weaker than $\neg P(\overline{I})$.

[6]In some of his philosophical notes, Gödel considered positiveness in the sense of "purely good" or "assertions" and suggested a simplification of the ontological argument without $A1$ [8, 435].

3 An S4 version of Gödel's argument with contrary terms

In his argument for the Cartesian lemma, Leibniz states that:

> [A]ll propositions which are necessarily true are either demonstrable or known per se. Therefore this proposition is not necessarily true, or it is not necessary that A and B should not be in the same subject. Therefore they can be in the same subject. [13, 167-168]

An understanding of necessity in terms of demonstrability can be found in many Leibnizian texts, and in some contexts it may also be associated with S4 modalities [1, 46-50].[7] We want to stress this connection also with respect to our pragmatic strategy to use the weakest possible logic as a basis for the proposed reconstruction. With regard to the proof of the Leibnizian lemma, it is enough to use proposed a specific instantiation of 5 stating that the possible necessary existence of God implies His necessary existence, and not 5 in its full extent. In this way we do not endorse the S5 system of modal logic, which seemed too strong to Gödel himself, and at the same time we are faithful to the original structure of the argument. This idea comes from [15].

The language of our approach is the same as that of Q2S5.

The logical axioms differ only for schema 5. Instead of this, we assume the weaker

(4) $\Box A \to \Box\Box A$

The rules are the same as in Q2S5. Now we are in Q2S4.

The admissibility of the necessitation rule in a weaker logic than S5 forces us to consider $\Box$ closures and $\Box\forall$ closures of the next axioms.

Concerning the cases of the comprehension schema, we accept $\Box\forall$ closures of

($G^{\leftrightarrow}$) $Gx \leftrightarrow \forall\varphi(\mathsf{P}(\varphi) \leftrightarrow \varphi x)$

(NE) $NEx \leftrightarrow \forall\varphi(\varphi Ess.x \to \Box\exists y\varphi y)$,

---

[7]S. Kovač stressed this connection with respect to the passage quoted above. He formalized it in a fragment of second order logic with a $\Diamond$ version of axiom 4: $\Diamond\Diamond A \to \Diamond A$ [11]. Accordnig to this approach, we begin with the assumption
(a1) $\Box A$ then (A is provable from other propositions or true per se).
Compossibility of two properties is defined as follows:
(**Comp**) $Comp(\lambda x.(Xx \wedge Yx)) \Leftrightarrow \Diamond\exists x(\lambda x.(Xx \wedge Yx))(x)$.
We consider two perfections X and Y. We assume indirectly that (1) $\neg Comp(\lambda x.(Xx \wedge Yx))$. Because (1) is not provable (X and Y are not *analyzable*) and not true per se, by (a1) we have (2) $\neg\Box\neg Comp(\lambda x.(Xx \wedge Yx))$. By $\Diamond/\Box$ and 2 we obtain (3) $\Diamond Comp(\lambda x.(Xx \wedge Yx))$ and so by $\Diamond$ version of 4 we obtain (4) $\Diamond\exists x(\lambda x.(Xx \wedge Yx))(x)$.

(again $\varphi Ess.x$ means $\varphi x \wedge \forall \psi(\psi(x) \rightarrow \Box \forall y(\varphi(y) \rightarrow \psi(y))))$

and the $\Box\forall$ closure of the implication

$(\stackrel{\rightarrow}{})$ $\varphi x \rightarrow \neg \overline{\varphi} x$

The following are the next specific axioms used in our formulation:

- $\Box\forall$ closures of

  $(A2)$ $\mathsf{P}(\varphi) \wedge \varphi \subset_\Box \psi \rightarrow \mathsf{P}(\psi)$ $(A4)$ $\mathsf{P}(\varphi) \rightarrow \Box \mathsf{P}(\varphi)$

- $\Box$ closure of

  $(\neg\mathsf{P})$ $\exists \varphi \neg \mathsf{P}(\varphi)$

- formulas:

  $(A3)$ $\mathsf{P}(G)$ $(A5)$ $\mathsf{P}(NE)$ $(A6)$ $\Diamond\Box\exists xGx \rightarrow \Box\exists xGx$

Our theory is the extension of **Q2S4** by the above axioms:

$$\mathsf{GO}^{\mathsf{P}}_{\beth_4} = \mathsf{Q2S4}[\Box\neg\mathsf{P}, A3, A5, A6, \Box\forall(\stackrel{\rightarrow}{}, \boldsymbol{G}^{\leftrightarrow}, \boldsymbol{NE}, A2, A4)].$$

In simple words, we developed an **S4** version of Gödel's ontological argument allowing contrary predicate terms.

The $\mathsf{GO}^{\mathsf{P}}_{\beth_4}$ is $\Box$ transparent like GO° (cf. ft. 3) The closures $\Box A3$ and $\Box A5$ are to be obtained from $A4$. We get $\Box A6$ from $\Box\neg\Box\exists xGx \vee \Box\exists xGx$ $[A6, \rightarrow /\vee, \Diamond/\Box]$ applied to $\Box A \vee \Box B \leftrightarrow \Box(\Box A \vee \Box B)$ [S4].

It is interesting that, unlike GO° our approach blocks full P-*necessitarianism* as expressed by the equivalence:

(NEC P) $\Diamond\mathsf{P}(\varphi) \leftrightarrow \mathsf{P}(\varphi) \leftrightarrow \Box\mathsf{P}(\varphi)$

In $\mathsf{GO}^{\mathsf{P}}_{\beth_4}$ we do not get the implication $\Diamond\mathsf{P}(\varphi) \rightarrow \mathsf{P}(\varphi)$, which is equivalent to $\neg\mathsf{P}(\varphi) \rightarrow \Box\neg\mathsf{P}(\varphi)$ added by Gödel to the axiom $A4$. Although, the formula $\Diamond\mathsf{P}(\varphi) \rightarrow \mathsf{P}(\varphi)$ is independently derivable in GO° via $A1$ or the instantiation of 5: $\Diamond\Box\mathsf{P}(\varphi) \rightarrow \Box\mathsf{P}(\varphi)$. However, we believe, once again that our approach is closer to a Leibnizian point of view. Let φ represent such a relative property as 'being the creator of a possible world w'. God may create any one of infinitely many worlds, but He chooses only one, and this is the actual world. By definition God possesses only positive properties, from which it follows that our property is possibly positive. But if $\Diamond\mathsf{P}(\varphi) \rightarrow \mathsf{P}(\varphi)$ is valid, then the above property relative to any possible world (including worlds that are not actual) is also positive. It follows from this that God, Who possesses all positive properties, is the creator of infinitely many possible worlds. This is obviously rejected by Leibniz.

The reconstruction of Gödel's argument is now simple. The formula Th1 is derivable in our theory in the same way as in fact 2.1. The Cartesian lemma **CL** follows directly from Th1 and $A3$. The proof of Th2 is formed as in fact 2.2. The Leibnizian lemma **LL** is derivable in the same way as in GO° (in step 7 of the proof we use our instantiation of 5, which we call $A6$). Finally the thesis Th3 is derived from **CL** and **LL**.

At the end of our consideration let us describe a model for $\mathsf{GO}^{\exists P}_4[\Box\forall \boldsymbol{I}_x]$ which falsifies the problematic implication $A1^\leftarrow$, the formula Th5 expressing $\Diamond\mathsf{P}$-*necessitarianism*, and also Th15 according to which all possible properties of God are His actual properties.

We adopt a modified semantics for $\mathsf{GO}^\circ[\Box\forall \boldsymbol{I}_x]$ given by Hájek [9]. A Kripke frame is $K = \langle W, R, D, Prop, \mathcal{P}\rangle$, where: W is a set of *possible worlds*; R is a reflexive and transitive *accessibility* relation in W; D is a set of *individuals* (all sets are nonempty); $Prop$ is a set of mappings $D \times W \to \{0, 1\}$ representing *properties* and such that:

(*) $\forall_{p \in Prop} \exists_{p' \in Prop} \forall_{d \in D} \forall_{w \in W} (p(d,w)) \cdot p'(d,w) = 0);$

$\mathcal{P}$ is a mapping that assigns value 1 to properties that are positive in possible worlds: $Prop \times W \to \{0, 1\}$, where

(**) $\forall_{p \in Prop} \forall_{w \in W} (\mathcal{P}(p, w) = 1 \implies \forall_{w'}(wRw' \implies \mathcal{P}(p, w') = 1)).$

The constant domain of individuals is not needed in case of our logic, however it simplifies our interpretation of $\mathsf{GO}^{\exists P}_4[\Box\forall \boldsymbol{I}_x]$.

We do not assume that the universal property is an element of *Prop*.
In Hájek's semantics, the characteristics of $\mathcal{P}$ do not depend on possible worlds.
The valuation function is $v(x) \in D$, $v(\tau) \in Prop$; for contrary terms we have:
$v(\tau) = p \implies \exists^1_{p'} v(\overline{\tau}) = p'$, where: $\forall_{d,w}(p(d,w) \cdot p'(d,w) = 0)$.

Taking valuation v, the validity of formulas in possible world $w \in W$ is described as follows:

$K, w \models^v x = y$ iff $v(x) = v(y)$
$K, w \models^v \varphi x$ iff $v(\varphi)(v(x), w) = 1$
$K, w \models^v \mathsf{P}(\varphi)$ iff $\mathcal{P}(v(\varphi), w) = 1$
$K, w \models^v \Box A$ iff $\forall_{w' \in W}(wRw' \implies K, w' \models^v A)$

The conditions for $\neg A$, $A \to B, \forall x A, \forall \varphi A$ are usual. The truth-conditions for G, and NE are already determined in the following way:

$K, w \models^v Gx$ iff $K, w \models^v \forall\varphi(P(\varphi) \leftrightarrow \varphi x)$
$K, w \models^v NEx$ iff $K, w \models^v \forall\varphi(\varphi Ess.x \to \Box\exists x \varphi x)$ (cf. (**Ess.**))

A formula A is valid in M iff $\forall_{w \in W} K, w \models^v A$, for all valuations v.

Similarly to Hájek, we assume that our structures fulfill the following conditions:

(C.I) $\forall x \exists \varphi \Box \forall y (\varphi y \leftrightarrow y = x)$ is valid

(C.NE) $\exists \varphi \Box \forall x (\varphi x \leftrightarrow \forall \psi (\psi Ess.x \to \Box \exists y \psi y))$ is valid

(C.G) $\exists \varphi \Box \forall x (\varphi x \leftrightarrow \forall \psi (P(\psi) \leftrightarrow \psi x))$

and that there is a $g \in D$ such that:

(g.1) $\forall_{p \in Prop} \forall_{w \in W} (p(g, w) = 1 \implies \forall_{w' \in W} (wRw' \implies p(g, w') = 1))$

(g.2) $\forall_{p \in Prop} (\mathcal{P}(p, w) = 1$ iff $\forall_{w' \in W} (wRw' \implies (p(g, w) = 1)))$

All theorems of $\mathsf{GO}_{\neg 4}^{\neg \mathsf{P}}[\boldsymbol{I}_x]$ are valid in the structures described above.

Now we consider $M = \langle \boldsymbol{K}, \boldsymbol{v} \rangle$ where $\boldsymbol{K} = <\boldsymbol{W}, \boldsymbol{R}, \boldsymbol{D}, \boldsymbol{Prop}, \boldsymbol{\mathcal{P}}>$;
$\boldsymbol{D} = \{g, a\}$; $\boldsymbol{W} = \{w_1, w_2\}$; $\boldsymbol{R} = \{<w_1, w_2>, <w_1, w_1>, <w_2, w_2>\}$;
$\boldsymbol{Prop} = \{G^*, NE^*, I_a^*, p^*, \emptyset^*, U^*\}$, where
$G^*(g, w_1) = G^*(g, w_2) = 1, G^*(a, w_1) = G^*(a, w_2) = 0; NE^* = G^*$;
$I_a^*(a, w_1) = I_a^*(a, w_2) = 1, I_a^*(g, w_1) = I_a^*(g, w_2) = 0$;
$p^*(g, w_1) = p^*(a, w_1) = 0, p^*(g, w_2) = p^*(a, w_2) = 1$;
$\forall_{d \in D} \forall_{w \in W} (\emptyset^*(d, w) = 0$ and $U^*(d, w) = 1)$;
$\boldsymbol{\mathcal{P}}(G^*, w_1) = \boldsymbol{\mathcal{P}}(G^*, w_2) = 1$;
$\boldsymbol{\mathcal{P}}(I_a^*, w_1) = \boldsymbol{\mathcal{P}}(I_a^*, w_2) = 0$;
$\boldsymbol{\mathcal{P}}(p^*, w_1) = 0, \boldsymbol{\mathcal{P}}(p^*, w_2) = 1$;
$\boldsymbol{\mathcal{P}}(\emptyset^*, w_1) = \boldsymbol{\mathcal{P}}(\emptyset^*, w_2) = 0$ and $\boldsymbol{\mathcal{P}}(U^*, w_1) = \boldsymbol{\mathcal{P}}(U^*, w_2) = 1$.

Let $\boldsymbol{v}(x) = g, \boldsymbol{v}(G) = G^*, \boldsymbol{v}(NE) = NE^*, \boldsymbol{v}(\varphi) = p^*, \boldsymbol{v}(\overline{\varphi}) = \emptyset^*$.

Now we can see that:

(*) $\boldsymbol{K}, w_1 \not\models^v \mathsf{P}(\varphi) \vee \mathsf{P}(\overline{\varphi})$.

Thus the axiom $A1^{\leftarrow}$ of GO° is falsified in our semantics.

The proposed formalization departs from Gödel's approach with respect to both the modalities in the strong sense of $\mathsf{S5}$ and classical term negation.

Despite the fact that our analysis brings Gödel's ideas close to Leibniz's view only to a limited extend, a few promising details can be highlited that capture the Leibnizian concept of God. Let us note that

(**) $\boldsymbol{K}, w_1 \not\models^v \Diamond \mathsf{P}(\varphi) \to \mathsf{P}(\varphi)$ (***) $\boldsymbol{K}, w_1 \not\models^v Gx \to (\Diamond \varphi x \to \varphi x)$

According to (**), our formalism rejects $\Diamond$P-*necessitarianism*. According to $G^{\leftrightarrow}$ and (*), it also rules out a conception of God as possessing every property or its negation.[8] Lastly, it can happened that some possible properties of God are not His actual properties (***), although all of His actual properties are necessarily possessed by Him (Th14 is provable in our theory).

References

[1] Adams, R. M. (1994), *Leibniz. Determinist, Theist, Idealist*, Oxford Univ. Press.

[2] Adams, R. M. (1995), "Introductory note to *1970". In: K. Gödel, *Collected Works*, vol. 3, 388-402.

[3] Anderson, C. A. (1990), "Some emendations of Gödel's ontological proof", *Faith and Philosophy*, 7, 291-303.

[4] Christian, C. (1989), "Gödel Version des Ontologischen Gottesbeweises", *Sitzungsberichte der Österreichischen Akademie der Wissenschaften*, Abt. II 198, 1-26.

[5] Czermak, J. (2002), "Abriss des ontologischen Argumentes". In: *Kurt Gödel. Wahrheit und Beweisbarkeit*, vol. II. *Kompedium zum Werk*, ed. B. Buldt, E. Köhler, M. Stöltzner, P. Weibel, C. Klein, W. DePauli-Schimanowich-Göttig, Viena: ÖBV et HPT VerlagsgmbH and Co. KG, 309-324.

[6] Futch, M. (2008), *Leibniz's Metaphysics of Time and Space*, Springer.

[7] Gödel, K. (1970), *Ontologischer Beweis*. February 10th 1970. Faksimile from Nachlaß reprinted in: *Kurt Gödel. Wahrheit und Beweisbarkeit*, vol. II. *Kompedium zum Werk*, ed. B. Buldt, E. Köhler, M. Stöltzner, P. Weibel, C. Klein, W. DePauli-Schimanowich-Göttig, Viena: ÖBV et HPT VerlagsgmbH and Co. KG, 307-308.

[8] Gödel, K. (1995), "Texts relating to the ontological proof". In: *Kurt Gödel, Collected Works*, ed. S. Feferman et al., vol. 3, Oxford: Oxford University Press, 429-437.

[9] Hájek, P. (2002), "Der Mathematiker und die Frage der Existenz Gottes (betreffend Gödels ontologischen Beweis)". In: *Kurt Gödel. Wahrheit und Beweisbarkeit*, vol. II. *Kompedium zum Werk*, ed. B. Buldt, E. Köhler, M. Stöltzner, P. Weibel, C. Klein, W. DePauli-Schimanowich-Göttig, Viena: ÖBV et HPT VerlagsgmbH and Co. KG, 325-336.

[10] Kovač, S. (2003), "Some weakened Gödelian ontological systems", *Journal of Philosophical Logic* 32, 565-588.

[11] Kovač, S. (2017), "The Concept of Possibility in Ontological Proofs", presentation of the contributed paper in 2nd World Congress on Logic and Religion University of Warsaw, 18.06-22.06.2017.

[12] Leibniz, G.W. (1987), *Sämtliche Schriften und Briefe*. Reihe II: *Philosophischer Briefwechsel*. Band 1. Auflage Darmstadt 1926; zweiter, unveränderter Nachdruck Berlin 1987. (Available on the Internet: http://www.uni-muenster.de/Leibniz/DatenII1/II1_B.pdf 11.09.2017).

---

[8]The Leibnizian God is neither finite nor infinite in time and space [6, 171-194].

[13] Leibniz, G.W. (1989), *Philosophical Papers and Letters*, ed. II, transl. and ed. by L. E. Loemker, The New Synthese Historical Library, vol. 2, Kluwer Academic Publishers.

[14] Perzanowski, J. (1989), *Logiki modalne a filozofia [Modal Logic and Philosophy]*, Jagiellonian Univ., Cracow.

[15] Świętorzecka, K. (2012), "Ontologiczny dowód Gödla z ograniczoną redukcją modalności" ["Gödel's ontological proof with limited reduction of modalities"], Przegląd Filozoficzny Nowa Seria, 3 (83), 21-34.

[16] Świetorzecka, K. (ed.) (2016), *Gödel's Ontological Argument. History, Modifications, and Controversies*, Semper.

A Case Study on Computational Hermeneutics: E. J. Lowe's Modal Ontological Argument

David Fuenmayor
Freie Universität Berlin, Germany
`david.fuenmayor@fu-berlin.de`

Christoph Benzmüller[*]
Freie Universität Berlin, Germany
University of Luxembourg, Luxembourg
`c.benzmueller@fu-berlin.de`

Abstract

Computers may help us to better understand (not just verify) arguments. In this article we defend this claim by showcasing the application of a new, computer-assisted interpretive method to an exemplary natural-language argument with strong ties to metaphysics and religion: E. J. Lowe's modern variant of St. Anselm's ontological argument for the existence of God. Our new method, which we call *computational hermeneutics*, has been particularly conceived for use in interactive-automated proof assistants. It aims at shedding light on the meanings of words and sentences by framing their inferential role in a given argument. By employing automated theorem reasoning technology within interactive proof assistants, we are able to drastically reduce (by several orders of magnitude) the time needed to test the logical validity of an argument's formalization. As a result, a new approach to logical analysis, inspired by Donald Davidson's account of radical interpretation, has been enabled. In computational hermeneutics, the utilization of automated reasoning tools effectively boosts our capacity to expose the assumptions we indirectly commit ourselves to every time we engage in rational argumentation and it fosters the explicitation and revision of our concepts and commitments.

---

[*]Funded by Volkswagen Foundation.

Part I: Introductory Matter

The traditional conception of logic as an *ars iudicandi* sees as its central role the classification of arguments into valid and invalid ones by identifying criteria that enable us to judge the correctness of (mostly deductive) inferences. However, logic can also be conceived as an *ars explicandi*, aiming at rendering the inferential rules implicit in our socio-linguistic argumentative praxis in a more orderly, more transparent, and less ambiguous way, thus setting the stage for an eventual critical assessment of our conceptual apparatus and inferential practices.

The novel approach we showcase in this article, called *computational hermeneutics*, is inspired by Donald Davidson's account of *radical interpretation* [18, 15]. It draws on the well-known *principle of charity* and on a holistic account of meaning, according to which the meaning of a term can only be given through the explicitation of the inferential role it plays in some theory (or argument) of our interest. We adopt the view that the process of logical analysis (aka. formalization) of a natural-language argument is itself a kind of interpretation, since it serves the purpose of *making explicit* the inferential relations between concepts and statements.[1] Moreover, the output of this process: *a* logical form, does not need to be unique, since it is dependent on a given background logical theory, or, as Davidson has put it:

"... much of the interest in logical form comes from an interest in logical geography: to give the logical form of a sentence is to give its logical location in the totality of sentences, to describe it in a way that explicitly determines what sentences it entails and what sentences it is entailed by. The location must be given relative to a specific deductive theory; so logical form itself is relative to a theory." ([16] p. 140)

Following the *principle of charity* while engaging in a process of logical analysis, requires us to search for plausible implicit premises, which would render the given argument as being logically valid and also foster important theoretical virtues such as consistency, non-circularity (avoiding 'question-begging'), simplicity, fruitfulness, etc. This task can be seen as a kind of conceptual *explication*.[2] In computational

---

[1] In recent times, this idea has become known as *logical expressivism* and has been championed, most notably, by the adherents of semantic inferentialism in the philosophy of language. Two paradigmatic book-length expositions of this philosophical position can be found in the works of Brandom [13] and Peregrin [37].

[2] Explication, in Carnap's sense, is a method of conceptual clarification, aimed at replacing an unclear 'fuzzy' pre-theoretical concept: an *explicandum*, by a new more exact concept with clearly defined rules of use: an *explicatum*, for use in a target theory. While Carnapian in spirit, our idea of explication focuses mostly on the activity of conceptual *explicitation* by the means of formal logic.

hermeneutics we carry it out by providing definitions (i.e. by directly relating a *definiendum* with a *definiens*) or by introducing formulas (e.g. as axioms) which relate a concept we are currently interested in (*explicandum*) with some other concepts which are themselves explicated in the same way (in the context of the same or some other background theory). The circularity involved in this process is an unavoidable characteristic of any interpretive endeavor and has been historically known as the *hermeneutic circle*. Thus, computational hermeneutics contemplates an iterative process of 'trial and error' where the adequacy of some newly introduced formula or definition becomes tested by computing, among others, the logical validity of the *whole* formalized argument. In order to explore the generally very wide space of possible formalizations (and also of interpretations) for even the simplest argument, we have to test its validity at least several hundreds of times (also to account for logical pluralism). It is here where the recent improvements and ongoing consolidation of modern automated theorem proving technology, in particular for higher-order logic (HOL), become handy. A concrete example of the application of this approach using the *Isabelle/HOL* [31] proof assistant to the logical analysis and interpretation of an ontological argument will be provided in the last section.

This article is divided in three parts. In the first one, we present the philosophical motivation and theoretical underpinnings of our approach; and we also outline the landscape of automated deduction. In the second part, we introduce the method of computational hermeneutics as an iterative process of conceptual explication. In the last part, we present our case study: the computer-assisted logical analysis and interpretation of E. J. Lowe's modal ontological argument, where our approach becomes exemplified.

Philosophical and Religious Arguments

Is it possible to find meaning in religious argumentation? Or is religion a conversation-stopper? Do religious beliefs provide a conceptual framework through which a believer's world-view is structured to such an extent, that the interpretation of religious arguments becomes a hopeless case? (given the apparent incommensurability between the conceptual schemes of speakers and interpreters of different creeds). The answer to these questions boils down to finding a way to acknowledge the variety of religious belief, while recognizing that we all share, at heart, a similar assortment of concepts and are thus able to understand each other. We argue for the role of logic as a common ground for understanding in general and, in particular, for theological argumentation. We reject therefore the view that deep religious convictions constitute an insurmountable obstacle for successful interreligious communication (e.g.

between believers and lay interpreters). Such views have been much discussed in religious studies. Terry Godlove, for instance, has convincingly argued in [24] against what he calls the "framework theory" in religious studies, according to which, for believers, religious beliefs shape the interpretation of most of the objects and situations in their lives. Here Godlove relies on Donald Davidson's rejection of "the very idea of a conceptual scheme" [17].

Davidson's criticism of what he calls "conceptual relativism" relies on the view that talk of incommensurable conceptual schemes is possible only on violating a correct understanding of interpretability, as developed in his theory of radical interpretation, especially vis-à-vis the well-known *principle of charity*. Furthermore, the kind of meaning holism implied by Davidson's account of interpretation suggests that we must share vastly more belief than not with anyone whose words and actions we are able to interpret. Thus, divergence in belief must be limited: If an interpreter is to interpret someone as asserting that Jerusalem is a holy place, she has to presume that the speaker holds true many closely related sentences; for instance, that Jerusalem is a city, that holy places are sites of pilgrimage, and, if the speaker is Christian, that Jesus is the son of God and lived in Jerusalem –and so on. Meaning holism requires us, so Godlove's thesis, to reject the notion that religions are alternative, incommensurable conceptual frameworks.

Drawing upon our experience with the computer-assisted reconstruction and assessment of ontological arguments for the existence of God [8, 9, 23, 7], we can bear witness to the previous claims. While looking for the most appropriate formalization of an argument, we have been led to consider further unstated assumptions (implicit premises) needed to reconstruct the argument as logically valid, and thus to ponder how much we may have departed from the original argument and to what extent we are still doing justice to the intentions of its author. We had to consider issues like the plausibility of our assumptions from the standpoint of the author and its compatibility with the author's purported beliefs (or what she said elsewhere).[3] Reflecting on this experience, we have become motivated to work out a computer-assisted interpretive approach drawing on semantic holism, which is especially suited for finding meaning in theological and metaphysical discourse.

We want to focus our inquiry on the issue of understanding a particular type of arguments and the role computers can play in it. We are thus urged to distinguish the kind of arguments we want to address from others that, on the one hand, rely

---

[3]More specifically, Eder and Ramharter [19] propose several criteria aimed at judging the adequacy of formal reconstructions of St. Anselm's ontological argument. They also show how such reconstructions help us gain a better understanding of this argument.

on appeals to faith and rhetorical effects, or, on the other hand, make use of already well-defined concepts with univocal usage, like in mathematics. We have already talked of religious arguments in the spirit of St. Anselm's ontological argument as some of the arguments we are interested in; we want, nonetheless, to generalize the domain of applicability of our approach to what we call 'philosophical' arguments (for lack of a better word), since we consider that many of the concepts introduced into these and many other kinds of philosophical discussions remain quite fuzzy and unclear ("explicanda" in Carnap's terminology). We want to defend the view that the process of *explicating* those philosophical concepts takes place in the very practice of argumentation through the *explicitation* of the inferential role they play in some theory or argument of our interest. In the context of a formalized argument (in some chosen logic), this task of conceptual explication can be carried out *systematically* by giving definitions or axiomatizing conceptual interrelations, and then using automated reasoning tools to explore the space of possible logical inferences. This approach, which we name computational hermeneutics, will be illustrated in the case study presented in the last section.

Top-down versus Bottom-up Approaches to Meaning

Above we have discussed the challenge of finding meaning in religious arguments. Determining meanings in philosophical contexts, however, has generally been considered a problematic task, especially when one wants to avoid the kind of ontological commitments resulting from postulating the existence, for every linguistic expression, of some obscure abstract being in need of definite identity criteria (cf. Quine's slogan "no entity without identity"). We want to talk here of the *meaning* of a linguistic expression (particularly of an argument) as *that* which the interpreter needs to grasp in order to *understand* it, and we will relate this to such blurry things as the inferential role of expressions.

In a similar vein, we also want to acknowledge the compositional character of natural and formalized languages, so we can think of the meaning of an argument as a function of the meanings of each of its constituent sentences (premises and conclusions) and their mode of combination (logical consequence relation).[4] Accordingly, we take the meaning of each sentence as resulting from the meaning of its constituent words

---

[4] Ideally, an argument would be analyzed as an island isolated from any external linguistic or pre-linguistic goings-on, to the extent that its validity would depend solely on what is explicitly stated (premises, inference rules, etc.); and, for instance, when *implicit* premises are brought to our attention, they should be made *explicit* and integrated into the argument accordingly –which must always remain an *intersubjectively* accessible artifact: a product of our socio-linguistic discursive practices. In the same spirit, it is also reasonable to expect of all sentences to derive their

(concepts) and their mode of combination. We can therefore, by virtue of compositionality, conceive a *bottom-up* approach for the interpretation of an argument, by starting with our pre-understanding (theoretical or colloquial) of its main concepts and then working our way up to an understanding of its sentences and their inferential interrelations.[5]

The bottom-up approach is the one usually employed in the formal verification of arguments (logic as *ars iudicandi*). However, it leaves open the question of how to arrive at the meaning of words beyond our initial pre-understanding of them. This question is central to our project, since we are interested in understanding (logic as *ars explicandi*) more than mere verifying. Thus, we want to complement the atomistic bottom-up approach with a holistic top-down one, by proposing a computer-supported method aimed at determining the meaning of expressions from their inferential role vis-à-vis argument's validity (which is determined for the argument *as a whole*), much in the spirit of Donald Davidson's program of *radical interpretation*.[6]

Radical Interpretation and the Principle of Charity

What is the use of radical interpretation in religious and metaphysical discourse? The answer is trivially stated by Davidson himself, who convincingly argues that "all understanding of the speech of another involves radical interpretation" ([15], p. 125). Furthermore, the impoverished evidential position we are faced with when interpreting metaphysical and theological arguments corresponds very closely to the starting situation Davidson contemplates in his thought experiments on *radical interpretation*, where he shows how an interpreter could come to understand someone's utterances without relying on any prior understanding of their language.[7]

---

meaning compositionally (in particular, we see no place for idioms in philosophical arguments). Unsurprisingly, these demands are never met in their entirety in real-world arguments.

[5]There is a well-known tension between the holistic nature of inferential roles and a compositional account of meaning. In computational hermeneutics, we aim at showing both approaches in action (top-down and bottom-up), thus demonstrating their compatibility in practice. For a theoretical treatment of the relationship between compositionality and meaning holism, we refer the reader to [35, 33, 34].

[6]The connections between Davidson's truth-centered theory of meaning and theories focusing on the inferential role of expressions (e.g. [13, 27, 12]) have been much discussed in the literature. While some authors (Davidson included) see both holistic approaches as essentially different, others (e.g. [45], [28], p. 72) have come to see Davidson's theory as an instance of inferential-role semantics. We side with the latter.

[7]For an interesting discussion of the relevance of Davidson's philosophy of language in religious studies, we refer the reader to [25].

Davidson's program builds on the idea of taking the notion of truth as basic and extracting from it an account of translation or interpretation satisfying two general requirements: (i) it must reveal the compositional structure of language, and (ii) it can be assessed using evidence available to the interpreter [15, 18].

The first requirement (i) is addressed by noting that a theory of truth in Tarski's style (modified to apply to natural language) can be used as a theory of interpretation. This implies that, for every sentence s of some object language L, a sentence of the form: «"s" is true in L iff p» (aka. T-schema) can be derived, where p acts as a translation of s into a sufficiently expressive metalanguage used for interpretation (note that in the T-schema the sentence p is being *used*, while s is only being *mentioned*). Thus, by virtue of the recursive nature of Tarski's definition of truth [43], the *compositional* structure of the object-language sentences becomes revealed.

From the point of view of computational hermeneutics, the sentence s is interpreted in the context of a given argument. The object language L thereby corresponds to the idiolect of the speaker (natural language plus some technical terms and background information), and the metalanguage is constituted by formulas of our chosen logic of formalization (some expressive logic XY) plus the turnstyle symbol $\vdash_{XY}$ signifying that an inference (argument) is valid in logic XY. As an illustration, consider the following instance of the T-schema used for some theological argument about monotheism: «"There is only one God" is true [in English, in the context of argument A] iff $A_1, A_2, ..., A_n \vdash_{HOL}$ "$\exists x.\ God\ x \land \forall y.\ God\ y \to y=x$"», where $A_1, A_2, ..., A_n$ correspond to the formalization of the premises of argument A and the turnstyle $\vdash_{HOL}$ corresponds to the standard logical consequence relation in higher-order logic (*HOL*). By comparing this with the T-schema («"s" is true in L iff p») we can notice that the *used* metalanguage sentence p can be paraphrased in the form: «"q" follows from the argument's premises [in HOL]» where the *mentioned* sentence q corresponds to the formalization (in some chosen logic) of the object sentence s. In this example we have considered a sentence playing the role of a conclusion which is being supported by some premises. It is however also possible to consider this same sentence in the role of a premise: «"There is only one God" is true [in the context of argument A] iff $A_1, A_2, ...,$ "$\exists x.\ God\ x \land \forall y.\ God\ y \to y=x$", $..., A_n \vdash_{HOL} C$»; now the truth of the sentence is postulated so that it can be used to validate C.[8] Most importantly, this example aims at illustrating how the interpretation of a sentence relates to its logical formalization and the inferential role it plays in a background

---

[8]We may actually want to weaken the double implication in this case, or work with an alternative notion of logical consequence. Moreover, other roles can be conceived for such a sentence in the context of an argument, for instance, it can also play the role of an unwanted conclusion: a sentence which we want to make sure it remains false no matter how we analyze the argument.

argument.

The second general requirement (ii) states that the interpreter has access to objective evidence in order to judge the appropriateness of her interpretations, i.e., access to the events and objects in the 'external world' that cause sentences to be true (or, in our case, arguments to be valid). In our approach, formal logic serves as a common ground for understanding. Computing the logical validity of a formalized argument constitutes the kind of objective (or, more appropriately, intersubjective) evidence needed to secure the adequacy of our interpretations, under the *charitable* assumption that the speaker follows (or at least accepts) similar logical rules as we do. In computational hermeneutics, the computer acts as an (arguably unbiased) arbiter deciding on the truth of a sentence in the context of an argument. In order to account for logical pluralism, computational hermeneutics targets the utilization of different kinds of classical and non-classical logics through the technique of *semantical embeddings* (see e.g. [6, 4]), which allows us to take advantage of the expressive power of classical higher-order logic (as a metalanguage) in order to embed the syntax and semantics of another logic (as an object language). Using the technique of semantical embeddings we can, for instance, embed a modal logic by defining the modal operators as meta-logical predicates. A framework for automated reasoning in different logics by applying the technique of semantical embeddings has been successfully implemented using automated theorem proving technology [21, 5].

Underlying his account of radical interpretation, there is a central notion in Davidson's theory: the *principle of charity*, which he holds as a condition for the possibility of engaging in any kind of interpretive endeavor. In a nutshell, the principle says that "we make maximum sense of the words and thoughts of others when we interpret in a way that optimizes agreement" [17]. The principle of charity builds on the possibility of intersubjective agreement about external facts among speaker and interpreter and can thus be invoked to make sense of a speaker's ambiguous utterances and, in our case, to presume (and foster) the validity of the argument we aim at interpreting. Consequently, in computational hermeneutics we assume from the outset that the argument's conclusions indeed follow from its premises and disregard formalizations that do not do justice to this postulate.

The Automated Reasoning Landscape

Automated reasoning is an umbrella term used for a wide range of technologies sharing the overall goal of mechanizing different forms of reasoning (understood as the ability to draw inferences). Born as a subfield of artificial intelligence with the aim of automatically generating mathematical proofs,[9] automated reasoning has moved to close proximity of logic and philosophy, thanks to substantial theoretical developments in the last decades. Nevertheless, its main field of application has mostly remained bounded to mathematics and hardware and software verification. In this respect, the field of *automated theorem proving* (ATP) has traditionally been its most developed subarea. ATP involves the design of algorithms that automate the process of construction (proof generation) and verification (proof checking) of mathematical proofs. Some extensive work has also been done in other non-deductive forms of reasoning (inductive, abductive, analogical, etc.). However, those fields remain largely underrepresented in comparison.

There have been major advances regarding the automatic generation of formal proofs during the last years, which we think make the utilization of formal methods in philosophy very promising and have even brought about some novel philosophical results (e.g. [9]). We will, on this occasion, restrain ourselves to the computer-supported interpretation of existing arguments, that is, to a situation where the given nodes/statements in the argument constitute a coarse grained "island proof structure" that needs to be rigorously assessed.

Proof checking can be carried out either non-interactively (for instance as a batch operation) or interactively by utilizing a proof assistant. A non-interactive proof-checking program would normally get as input some formula (string of characters in some predefined syntax) and a context (some collection of such formulas) and will, in positive cases, generate a listing of the formulas (in the given context) from which the input formula logically follows, together with the name of the proof method[10] used and, in some cases, a proof string (as in the case of proof generators). Some proof checking programs, called *model finders*, are specialized in searching for models and, more importantly, countermodels for a given formula. This functionality proves very useful in practice by sparing us the thankless task of trying to prove non-theorems.

Human guidance is oftentimes required by theorem provers in order to effectively solve interesting problems. A need has been recognized for the synergistic combination of the vast memory resources and information-processing capabilities of modern computers, together with human ingenuity, by allowing people to give hints to these tools by the means of especially crafted user interfaces. The field of *interactive the-*

---

[9]For instance, the first widely recognized AI system: *Logic Theorist*, was able to prove 38 of the first 52 theorems of Whitehead and Russell's "Principia Mathematica" back in 1956.

[10]For instance, some of the proof methods commonly employed by the *Isabelle/HOL* proof assistant are: term rewriting, classical reasoning, tableaus, model elimination, ordered resolution and paramodulation.

orem proving has grown out of this endeavor and its software programs are known as *proof assistants*.[11]

Automated reasoning is currently being applied to solve problems in formal logic, mathematics and computer science, software and hardware verification and many others. For instance, the Mizar Library[12] and TPTP (Thousands of Problems for Theorem Provers) [42] are two of the biggest libraries of such problems being maintained and updated on a regular basis. There is also a yearly competition among automated theorem provers held at the CADE conference [36], whose problems are selected from the TPTP library.

Automated theorem provers (particularly focusing on higher order logics) have been used to assist in the formalization of many advanced mathematical proofs such as Erdös-Selberg's proof of the *Prime Number Theorem* (about 30,000 lines in Isabelle), the proof of the *Four Color Theorem* (60,000 lines in Coq), and the proof of the *Jordan Curve Theorem* (75,000 lines in HOL-Light) [40]. The monumental proof of Kepler's conjecture by Thomas Hales and his research team has been recently formalized and verified using the HOL-Light and Isabelle proof assistants as part of the *Flyspeck project* [26].

Isabelle [31] is the proof assistant we will use to illustrate our *computational hermeneutics* method. Isabelle offers a structured proof language called *Isar* specifically tailored for writing proofs that are both computer- and human-readable and which focuses on higher-order classical logic. The different variants of the ontological argument assessed in our case study are formalized directly in Isabelle's HOL dialect or, for the modal variants, through the technique of shallow semantical embeddings [6].

Part II: The Computational Hermeneutics Method

It is easy to argue that using computers for the assessment of arguments brings us many *quantitative* advantages, since it gives us the means to construct and verify proofs easier, faster, and much more reliably. Furthermore, a main task of this paper is to illustrate a central *qualitative* advantage of computer-assisted argumentation: It enables a different, *holistic* approach to philosophical argumentation.

---

[11] A survey and system comparison of the most famous interactive proof assistants has been carried out in [44]. The results of this survey remain largely accurate to date.

[12] *Mizar* proofs and their corresponding articles are published regularly in the peer-reviewed *Journal of Formalized Mathematics*.

Holistic Approach: Why Feasible Now?

Let us imagine the following scenario: A philosopher working on a formal argument wants to test a variation on one of its premises or definitions and find out if the argument still holds. Our philosopher is working with pen and paper and she follows some chosen proof procedure (e.g. natural deduction or sequent calculus). Depending on her calculation skills, this may take some minutes, if not much longer, to be carried out. It seems clear that she cannot allow herself many of such experiments on such conditions.

Now compare the above scenario to another one in which our working philosopher can carry out such an experiment in just a few seconds and with no effort, by employing an automated theorem prover. In a best-case scenario, the proof assistant would automatically generate a proof (or the sketch of a countermodel), so she just needs to interpret the results and use them to inform her new conjectures. In any case, she would at least know if her speculations had the intended consequences, or not. After some minutes of work, she will have tried plenty of different variations of the argument while getting real-time feedback regarding their suitability.[13]

We aim at showing how this radical *quantitative* increase in productivity does indeed entail a *qualitative* change in the way we approach formal argumentation, since it allows us to take things to a whole new level (note that we are talking here of many hundreds of such trial-and-error 'experiments' that would take weeks or months if using pen and paper). Most importantly, this qualitative leap opens the door for the possibility of automating the process of logical analysis for natural-language arguments with regard to their subsequent computer-assisted critical evaluation.

The Approach

Computational hermeneutics is a holistic iterative enterprise, where we evaluate the adequacy of some candidate formalization of a sentence by computing the logical validity of the whole argument. We start with formalizations of some simple statements (taking them as tentative) and use them as stepping stones on the way to the formalization of other argument's sentences, repeating the procedure until arriving at a state of *reflective equilibrium*: A state where our beliefs and commitments have

---

[13]The situation is obviously idealized, since, as is well known, most of theorem-proving problems are computationally complex and even undecidable, so in many cases a solution will take several minutes or just never be found. Nevertheless, as work in the emerging field of *computational metaphysics* [32, 1, 41, 8, 9] suggests, the lucky situation depicted above is not rare.

the highest degree of coherence and acceptability.[14] In computational hermeneutics, we work iteratively on an argument by temporarily fixing truth-values and inferential relations among its sentences, and then, after choosing a logic for formalization, working back and forth on the formalization of its premises and conclusions by making gradual adjustments while getting automatic feedback about the suitability of our speculations. In this fashion, by engaging in a dialectic questions-and-answers ('trial-and-error') interaction with the computer, we work our way towards a proper understanding of an argument by circular movements between its parts and the whole (hermeneutic circle).

A rough outline of the iterative structure of the *computational hermeneutics* approach is as follows:

1. **Argument reconstruction** (initially in natural language):

    a. **Add or remove sentences and choose their truth-values.**
       Premises and desired conclusions would need to become true, while some other 'unwanted' conclusions would have to become false. Deciding on these issues expectedly involves a fair amount of human judgment.

    b. **Establish inferential relations,** i.e., determine the extension of the logical consequence relation: which sentences should follow (logically) from which others. This task can be done manually or automatically by letting our automated tools find this out for themselves, provided the logic for formalization has been selected and argument has already been roughly formalized (hence the mechanization of this step becomes feasible only after at least one outermost iteration). Automating this task frequently leads to the simplification of the argument, since current theorem provers are quite good at detecting idle axioms (see e.g. Isabelle's *Sledgehammer* tool [10]).

2. **Selection of a logic for formalization,** guided by determining the logical structure of the natural-language sentences occurring in the argument. This task can be partially automated (using the *semantical embeddings* technique) by searching a catalog of different embedded logics (in HOL) and selecting a

---

[14]We have been inspired by John Rawls' notion of *reflective equilibrium* as a state of balance or coherence between a set of general principles and particular judgments (where the latter follow from the former). We arrive at such a state through a deliberative give-and-take process of mutual adjustment between principles and judgments. More recent methodical accounts of reflective equilibrium have been proposed as a justification condition for scientific theories [20] and objectual understanding [2], and also as an approach to logical analysis [39].

candidate logic (modal, free, deontic, etc.) satisfying some particular syntactic or semantic criteria.

3. **Argument formalization (in the chosen logic),** while getting continuous feedback from our automated reasoning tools about the argument's correctness (validity, consistency, non-circularity, etc.). This stage is itself iterative, since, for every sentence, we charitably (in the spirit of the *principle of charity*) try several different formalizations until getting a correct argument. Here is where we take most advantage of the real-time feedback offered by our automated tools. Some main tasks to be considered are:

   a. **Translate natural-language sentences into the target logic,** by relying either on our pre-understanding or on provided definitions of the argument's terms.

   b. **Vary the logical form of already formalized sentences.** This can be done systematically and automatically by relying upon a catalog of (consistent) logical variations of formulas (see *semantical embeddings*) and the output of automated tools (ATPs, model finders, etc.).

   c. **Bring related terms together,** either by introducing definitions or by axiomatizing new interrelations among them. These newly introduced formulas can be translated back into natural language to be integrated into the argument in step (1.a), thus being disclosed as former *implicit* premises. The process of searching for additional premises with the aim of rendering an argument formally correct can be seen as a kind of abductive reasoning ('inference to the best explanation') and thus needs human support (at least at the current state of the art).

4. **Are termination criteria satisfied?** That is, have we arrived at a state of *reflective equilibrium*? If not, we would come back to some early stage. Termination criteria can be derived from the adequacy criteria of formalization found in the literature on logical analysis (see e.g. [3, 14, 38, 39]). An equilibrium may be found after several iterations without any significant improvements.[15]

---

[15] In particular, inferential adequacy criteria lend themselves to the application of automated deduction tools. Consider, for instance, Peregrin and Svoboda's [39] proposed criteria:
(i) The *principle of reliability*: "ϕ counts as an adequate formalization of the sentence S in the logical system L only if the following holds: If an argument form in which ϕ occurs as a premise or as the conclusion is valid in L, then all its perspicuous natural language instances in which S appears as a natural language instance of ϕ are intuitively correct arguments."
(ii) The *principle of ambitiousness*: "ϕ is the more adequate formalization of the sentence S in

Furthermore, the introduction of automated reasoning and linguistic analysis tools makes it feasible to apply these criteria to compute, in seconds, the degree of 'fitness' of some candidate formalization for a sentence (in the context of an argument).

Part III: Lowe's Modal Ontological Argument

In this section, the main contribution of this article, we illustrate the computer-supported interpretation of a variant of St. Anselm's ontological argument for the existence of God, using *Isabelle/HOL*.[16] This argument, which was introduced by the philosopher E. J. Lowe in an article named "A Modal Version of the Ontological Argument" [30], serves here as an exemplary case for an interesting and sufficiently complex, systematic argument with strong ties to metaphysics and religion. The interpretation of Lowe's argument thus makes for an ideal showcase for *computational hermeneutics* in practice.

Lowe offers in his article a new modal variant of the ontological argument, which is specifically aimed at proving the *necessary* existence of God. In a nutshell, Lowe's argument works by first postulating the existence of *necessary abstract* beings, i.e., abstract beings that exist in every possible world (e.g. numbers). He then introduces the concepts of *ontological dependence* and *metaphysical explanation* and argues that the existence of every (mind-dependent) abstract being is ultimately explained by some concrete being (e.g. a mind). By interrelating the concepts of *dependence* and *explanation*, he argues that the concrete being(s), on which each necessary abstract being depends for its existence, must also be necessary. This way he proves the existence of at least one *necessary concrete* being (i.e. God, according to his definition).

Lowe further argues that his argument qualifies as a modal ontological argument, since it focuses on *necessary* existence, and not just existence of some kind of supreme being. His argument differs from other familiar variants of the modal ontological argument (like Gödel's) in that it does not appeal, in the first place, to the possible existence of God in order to use the modal *S5* axioms to deduce its necessary ex-

---

the logical system L the more natural language arguments in which S occurs as a premise or as the conclusion, which fall into the intended scope of L and which are intuitively perspicuous and correct, are instances of valid argument forms of S in which ϕ appears as the formalization of S." ([39] pp. 70-71).

[16] We refer the reader to [22] for further details. That computer-verified article has been completely written in the Isabelle proof assistant and thus requires some familiarity with this system.

istence as a conclusion.[17] Lowe wants therefore to circumvent the usual criticisms to the *S5* axiom system, like implying the unintuitive assertion that whatever is possibly necessarily the case is thereby actually the case.

The structure of Lowe's argument is very representative of methodical philosophical arguments. It features eight premises from which new inferences are drawn until arriving at a final conclusion: the necessary existence of God (which in this case amounts to the existence of some necessary concrete being). The argument's premises are reproduced verbatim below:

(P1) God is, by definition, a necessary concrete being.

(P2) Some necessary abstract beings exist.

(P3) All abstract beings are dependent beings.

(P4) All dependent beings depend for their existence on independent beings.

(P5) No contingent being can explain the existence of a necessary being.

(P6) The existence of any dependent being needs to be explained.

(P7) Dependent beings of any kind cannot explain their own existence.

(P8) The existence of dependent beings can only be explained by beings on which they depend for their existence.

We will consider here only a representative subset of the argument's conclusions, which are reproduced below:

(C1) All abstract beings depend for their existence on concrete beings. (Follows from P3 and P4 together with definitions D3 and D4.)

(C5) In every possible world there exist concrete beings. (Follows from C1 and P2.)

(C7) The existence of necessary abstract beings needs to be explained. (Follows from P2, P3 and P6.)

(C8) The existence of necessary abstract beings can only be explained by concrete beings. (Follows from C1, P3, P7 and P8.)

(C9) The existence of necessary abstract beings is explained by one or more necessary concrete beings. (Follows from C7, C8 and P5.)

---

[17] As shown in [8], modal logic *KB* actually suffices to prove Scott's variant of Gödel's argument; this was probably not known to Lowe though.

(C10) A necessary concrete being exists. (Follows from C9.)

Lowe also introduces some informal definitions which should help the reader to understand some of the concepts involved in his argument (necessity, concreteness, ontological dependence, metaphysical explanation, etc.). In the following discussion, we will see that most of these definitions do not bear the significance Lowe claims.

(D1) x is a necessary being := x exists in every possible world.

(D2) x is a contingent being := x exists in some but not every possible world.

(D3) x is a concrete being := x exists in space and time, or at least in time.

(D4) x is an abstract being := x does not exist in space or time.

(D5) x depends for its existence on y := necessarily, x exists only if y exists.

In the following sections we use computational hermeneutics to interpret iteratively the argument shown above (by reconstructing it formally in different variations and in different logics). We compile in each section the results of a series of iterations and present them as a new variant of the original argument. We want to illustrate how the argument (as well as our understanding of it) gradually evolves as we experiment with different combinations of definitions, premises and logics for formalization.

First Iteration Series: Initial Formalization

Let us first turn to the formalization of premise P1: "God is, by definition, a necessary concrete being".[18]

In order to shed light on the concept of *necessariness* (i.e. being a necessary being) employed in this argument, we have a look at the definitions D1 and D2 provided by the author. They relate the concepts of necessariness and contingency (i.e. being a contingent being) with existence:[19]

---

[18] When the author says of something that it is a "necessary concrete being" we will take him to say that it is both necessary and concrete. Certainly, when we say of Tom that he is a lousy actor, we just don't mean that he is lousy and that he also acts. For the time being, we won't differentiate between predicative and attributive uses of adjectives, so we will formalize both sorts as unary predicates; since the particular linguistic issues concerning attributive adjectives don't seem to play a role in this argument. In the spirit of the *principle of charity*, we may justify adding further complexity to the argument's formalization if we later find out that it is required for its validity.

[19] Here, the concepts of necessariness and contingency are meant as properties of beings, in contrast to the concepts of necessity and possibility which are modals. We will see later how both pairs of concepts can be related in order to validate this argument.

(D1) *x is a necessary being* := *x exists in every possible world.*

(D2) *x is a contingent being* := *x exists in some but not every possible world.*

The two definitions above, aimed at explicating the concepts of necessariness and contingency by reducing them to existence and quantification over possible worlds, have a direct impact on the choice of a logic for formalization. They not only call for some kind of modal logic with possible-world semantics but also lead us to consider the complex issue of existence, since we need to restrict the domain of quantification at every world.

The choice of a modal logic for formalization has brought to the foreground an interesting technical constraint: The Isabelle proof assistant (as well as others) does not natively support modal logics. We have used, therefore, a technique known as *semantical embedding*, which allows us to take advantage of the expressive power of higher-order logic in order to embed the syntax and semantics of an object language. Here we draw on previous work on the embedding of multimodal logics in HOL [6], which has successfully been applied to the analysis and verification of ontological arguments (e.g. [9, 8, 7, 23]). Using this technique, we can embed a modal logic K by defining the $\Box$ and $\Diamond$ operators using restricted quantification over the set of *reachable* worlds (using a *reachability relation* R as a guard). Note that, in the following definitions, the type *wo* is declared as an abbreviation for $w \Rightarrow bool$, which corresponds to the type of a function mapping worlds (of type w) to boolean values. *wo* thus corresponds to the type of a world-dependent formula (i.e. its *truth set*).

consts $R::w \Rightarrow w \Rightarrow bool$ (**infix R**) — Reachability relation
abbreviation *mbox* :: $wo \Rightarrow wo$ ($\Box$-)
  where $\Box \varphi \equiv \lambda w. \forall v. (w \text{ R } v) \longrightarrow (\varphi \ v)$
abbreviation *mdia* :: $wo \Rightarrow wo$ ($\Diamond$-)
  where $\Diamond \varphi \equiv \lambda w. \exists v. (w \text{ R } v) \wedge (\varphi \ v)$

The 'lifting' of the standard logical connectives to type *wo* is straightforward. Validity is consequently defined as truth in *all* worlds and represented by wrapping the formula in special brackets ($\lfloor - \rfloor$).

abbreviation *valid*::$wo \Rightarrow bool$ ($\lfloor - \rfloor$) **where** $\lfloor \psi \rfloor \equiv \forall w. (\psi \ w)$

We verify our embedding by using Isabelle's simplifier to prove the K principle and the *necessitation* rule.

lemma K: $\lfloor (\Box(\varphi \rightarrow \psi)) \rightarrow (\Box \varphi \rightarrow \Box \psi) \rfloor$ **by** *simp* — Verifying K principle
lemma *NEC*: $\lfloor \varphi \rfloor \Longrightarrow \lfloor \Box \varphi \rfloor$ **by** *simp*   — Verifying *necessitation* rule

Regarding existence, we need to commit ourselves to a certain position in metaphysics known as *metaphysical contingentism*, which roughly states that the existence of any entity is a contingent fact: some entities can exist at some worlds, while not existing at some others. The negation of metaphysical contingentism is known as *metaphysical necessitism*, which basically says that all entities must exist at all possible worlds. By not assuming contingentism and, therefore, assuming necessitism, the whole argument would become trivial, since all beings would end up being trivially necessary (i.e. existing in all worlds).[20]

We hence restrict our quantifiers so that they range only over those entities that 'exist' (i.e. are actualized) at a given world. This approach is known as *actualist quantification* and is implemented, using the semantical embedding technique, by defining a world-dependent meta-logical 'existence' predicate (called "actualizedAt" below), which is the one used as a guard in the definition of the quantifiers. Note that the type e characterizes the domain of all beings (i.e. existing and non-existing entities), and the type wo characterizes sets of worlds. The term "isActualized" thus relates beings to worlds.

consts *isActualized*::$e \Rightarrow wo$ (**infix** *actualizedAt*)

abbreviation *forallAct*::$(e \Rightarrow wo) \Rightarrow wo$ ($\forall^A$)
  where $\forall^A \Phi \equiv \lambda w. \forall x. (x\ actualizedAt\ w) \longrightarrow (\Phi\ x\ w)$
abbreviation *existsAct*::$(e \Rightarrow wo) \Rightarrow wo$ ($\exists^A$)
  where $\exists^A \Phi \equiv \lambda w. \exists x. (x\ actualizedAt\ w) \land (\Phi\ x\ w)$

The corresponding binder syntax is defined below.

abbreviation *mforallActB*::$(e \Rightarrow wo) \Rightarrow wo$ (**binder**$\forall^A[8]9$)
  where $\forall^A x. (\varphi\ x) \equiv \forall^A \varphi$
abbreviation *mexistsActB*::$(e \Rightarrow wo) \Rightarrow wo$ (**binder**$\exists^A[8]9$)
  where $\exists^A x. (\varphi\ x) \equiv \exists^A \varphi$

We use a model finder (Isabelle's Nitpick tool [11]) to verify that actualist quantification validates neither the Barcan formula nor its converse. For the conjectured lemma, Nitpick finds a countermodel, i.e. a model (satisfying all axioms) which falsifies the given formula. The formula is consequently non-valid (as indicated by the Isabelle's "oops" keyword).

lemma $\lfloor (\forall^A x.\ \Box(\varphi\ x)) \rightarrow \Box(\forall^A x.\ \varphi\ x) \rfloor$
  nitpick oops — Countermodel found: formula not valid
lemma $\lfloor \Box(\forall^A x.\ \varphi\ x) \rightarrow (\forall^A x.\ \Box(\varphi\ x)) \rfloor$
  nitpick oops — Countermodel found: formula not valid

---

[20]Metaphysical contingentism looks *prima facie* like a very natural assumption to make; nevertheless an interesting philosophical debate between advocates of necessitism and contingentism has arisen during the last years, especially in the wake of Timothy Williamson's work on the metaphysics of modality (see [46]).

Unrestricted (aka. possibilist) quantifiers, in contrast, validate both the Barcan formula and its converse.

lemma $\lfloor(\forall x.\Box(\varphi\ x)) \rightarrow \Box(\forall x.(\varphi\ x))\rfloor$
  by *simp* — Proven by Isabelle's simplifier
lemma $\lfloor\Box(\forall x.(\varphi\ x)) \rightarrow (\forall x.\Box(\varphi\ x))\rfloor$
  by *simp* — Proven by Isabelle's simplifier

With actualist quantification in place we can: (i) the concept of existence becomes formalized (explicated) in the usual form by using a restricted particular quantifier ($\approx$ stands for the unrestricted identity relation on all objects), (ii) necessariness becomes formalized as existing necessarily, and (iii) contingency becomes formalized as existing possibly but not necessarily.

definition *Existence*::$e \Rightarrow wo$ (*E!*) **where** $E!\ x \equiv \exists^A y.\ y \approx x$

definition *Necessary*::$e \Rightarrow wo$ **where** *Necessary* $x \equiv \Box E!\ x$
definition *Contingent*::$e \Rightarrow wo$ **where** *Contingent* $x \equiv \Diamond E!\ x \land \neg$*Necessary* x

Note that we have just chosen a logic for formalization: a free quantified modal logic *K* with positive semantics. The logic is *free* because the domain of quantification (for actualist quantifiers) is a proper subset of our universe of discourse (so we can refer to non-existing objects). The semantics is *positive* because we have placed no restriction regarding predication on non-existing objects, so they are also allowed to exemplify properties and relations. We are also in a position to embed stronger normal modal logics (*KB, KB5, S4, S5, etc.*) by restricting the reachability relation *R* with additional axioms, if needed.

Having chosen our logic, we can now turn to the formalization of the concepts of abstractness and concreteness. As seen previously, Lowe has already provided us with an explication of these concepts:

(D3) *x is a concrete being := x exists in space and time, or at least in time.*

(D4) *x is an abstract being := x does not exist in space or time.*

Lowe himself acknowledges that the explication of these concepts in terms of existence "in space and time" is superfluous, since we are only interested in them being complementary.[21] Thus, we start by formalizing concreteness as a *primitive* world-dependent predicate and then derive abstractness from it, namely as its negation.

---

[21] We quote from Lowe's original article: "Observe that, according to these definitions, a being cannot be both concrete and abstract: being concrete and being abstract are mutually exclusive properties of beings. Also, all beings are either concrete or abstract ... the abstract/concrete distinction is exhaustive. Consequently, a being is concrete if and only if it is not abstract."

consts *Concrete*::$e\Rightarrow wo$
abbreviation *Abstract*::$e\Rightarrow wo$ **where** *Abstract* $x \equiv \neg(\textit{Concrete } x)$

We can now formalize the definition of Godlikeness (P1) as follows:

abbreviation *Godlike*::$e\Rightarrow wo$ **where** *Godlike* $x \equiv \textit{Necessary } x \wedge \textit{Concrete } x$

We also formalize premise P2 ("Some necessary abstract beings exist") as shown below:

axiomatization where
P2: $\lfloor \exists^A x.\ \textit{Necessary } x \wedge \textit{Abstract } x \rfloor$

Let us now turn to premises P3 ("All abstract beings are dependent beings") and P4 ("All dependent beings depend for their existence on independent beings"). We have here three new terms to be explicated: two predicates "dependent" and "independent" and a relation "depends (for its existence) on", which has been called *ontological dependence* by Lowe. Following our linguistic intuitions concerning their interrelation, we start by proposing the following formalization:

consts *dependence*::$e\Rightarrow e\Rightarrow wo$ (**infix** *dependsOn*)
definition *Dependent*::$e\Rightarrow wo$ **where** *Dependent* $x \equiv \exists^A y.\ x \textit{ dependsOn } y$
abbreviation *Independent*::$e\Rightarrow wo$ **where** *Independent* $x \equiv \neg(\textit{Dependent } x)$

We have formalized ontological dependence as a *primitive* world-dependent relation and refrained from any explication (as suggested by Lowe).[22] Moreover, an entity is *dependent* if and only if there *actually exists* an object y such that x *depends for its existence* on it; accordingly, we have called an entity *independent* if and only if it is not dependent.

As a consequence, premises P3 ("All abstract beings are dependent beings") and P4 ("All dependent beings depend for their existence on independent beings") become formalized as follows.

axiomatization where
P3: $\lfloor \forall^A x.\ \textit{Abstract } x \rightarrow \textit{Dependent } x \rfloor$ **and**

---

[22] An explication of this concept has been suggested by Lowe in definition D5 ("x depends for its existence on y := necessarily, x exists only if y exists"). Concerning this alleged definition, he has written in a footnote to the same article: "Note, however, that the two definitions (D5) and (D6) presented below are not in fact formally called upon in the version of the ontological argument that I am now developing, so that in the remainder of this chapter the notion of existential dependence may, for all intents and purposes, be taken as primitive. There is an advantage in this, inasmuch as finding a perfectly apt definition of existential dependence is no easy task, as I explain in 'Ontological Dependence.'" Lowe refers hereby to his article on ontological dependence in the *Stanford Encyclopedia of Philosophy* [29] for further discussion.

P4: $\lfloor \forall^A x.\ \textit{Dependent } x \rightarrow (\exists^A y.\ \textit{Independent } y \wedge x\ \textit{dependsOn } y)\rfloor$

Concerning premises P5 ("No contingent being can explain the existence of a necessary being") and P6 ("The existence of any dependent being needs to be explained"), a suitable formalization for expressions of the form: "the entity X explains the existence of Y" and "the existence of X is explained" needs to be found.[23] These expressions rely on a single binary relation, which will initially be taken as *primitive*. This relation has been called *metaphysical explanation* by Lowe.[24]

consts *explanation*::$e \Rightarrow e \Rightarrow wo$ (**infix** *explains*)
definition *Explained*::$e \Rightarrow wo$ **where** *Explained* $x \equiv \exists^A y.\ y\ \textit{explains } x$

axiomatization where
P5: $\lfloor \neg(\exists^A x.\ \exists^A y.\ \textit{Contingent } y \wedge \textit{Necessary } x \wedge y\ \textit{explains } x)\rfloor$

Premise P6, together with the last two premises: P7 ("Dependent beings of any kind cannot explain their own existence") and P8 ("The existence of dependent beings can only be explained by beings on which they depend for their existence"), were introduced by Lowe in order to relate the concept of *metaphysical explanation* to *ontological dependence*.[25]

axiomatization where
P6: $\lfloor \forall x.\ \textit{Dependent } x \rightarrow \textit{Explained } x \rfloor$ **and**
P7: $\lfloor \forall x.\ \textit{Dependent } x \rightarrow \neg(x\ \textit{explains } x)\rfloor$ **and**
P8: $\lfloor \forall x\ y.\ y\ \textit{explains } x \rightarrow x\ \textit{dependsOn } y \rfloor$

Although the last three premises seem to couple very tightly the concepts of (metaphysical) explanation and (ontological) dependence, both concepts are not meant by the author to be equivalent.[26] We have used Nitpick to test this claim. Since a countermodel has been found, we have proven that the inverse equivalence of metaphysical explanation and ontological dependence is not implied by the axioms (a

---

[23] Note that we have omitted the expressions "can" and "needs to" in our formalization, since they seem to play here only a rhetorical role. As in the case of attributive adjectives discussed before, we first aim at the simplest workable formalization; however, we are willing to later improve on this formalization in order to foster argument's validity, in accordance to the *principle of charity*.

[24] This concept is closely related to what has been called *metaphysical grounding* in contemporary literature.

[25] Note that we use non-restricted quantifiers for the formalization of the last three premises in order to test the argument's validity under the strongest assumptions. As before, we turn a blind eye to the modal expression "can".

[26] Lowe says: "Existence-explanation is not simply the inverse of existential dependence. If x depends for its existence on y, this only means that x cannot exist without y existing. This is not at all the same as saying that x exists because y exists, or that x exists in virtue of the fact that y exists."

screenshot showing Nitpick's text-based representation of such a model is provided below).

lemma $\lfloor \forall x\, y.\ x\ explains\ y \leftrightarrow y\ dependsOn\ x \rfloor$ **nitpick**[*user-axioms*] **oops**

For any being, however, having its existence "explained" is equivalent to its existence being "dependent" (on some other being). This follows already from premises P6 and P8, as shown above by Isabelle's prover.

lemma $\lfloor \forall x.\ Explained\ x \leftrightarrow Dependent\ x \rfloor$
  using *P6 P8 Dependent-def Explained-def* **by** *auto*

The Nitpick model finder is also useful to check axioms' consistency at any stage during the formalization of an argument. We instruct Nitpick to search for a model satisfying some tautological sentence (here we use a trivial 'True' proposition), thus demonstrating the satisfiability of the argument's axioms. Nitpick's output is a text-based representation of the found model (or a message indicating that no model, up to a predefined cardinality, could be found). This information is very useful to inform our future decisions. The screenshot below (taken from the Isabelle proof assistant) shows the model found by Nitpick, which satisfies the argument's formalized premises:

lemma *True* **nitpick**[*satisfy, user-axioms*] **oops**

```
206 lemma "⌊x explains y ↔ y dependsOn x⌋" nitpick[user_axioms] oops
207
208 lemma True nitpick[satisfy, user_axioms] oops
```

Nitpicking formula...
Nitpick found a model for card e = 2 and card w = 1:

  Constants:
    op actualizedAt = (λx. _)((a, w_1) := True, (b, w_1) := True)
    op r = (λx. _)((w_1, w_1) := False)
    Concrete = (λx. _)((a, w_1) := False, (b, w_1) := True)
    op dependsOn =
      (λx. _)
      ((a, a, w_1) := True, (a, b, w_1) := True,
       (b, a, w_1) := False, (b, b, w_1) := False)
    op explains =
      (λx. _)
      ((a, a, w_1) := False, (a, b, w_1) := False,
       (b, a, w_1) := True, (b, b, w_1) := False)

In this case, Nitpick was able to find a model satisfying the given tautology; this means that all axioms defined so far are consistent. The model found consists of two individual objects a and b and and a single world w_1, which is not connected via the reachability relation R to itself. We furthermore have in world w_1: b is concrete, a is not; a depends on b and itself, while b depends on no other object; b is the only object that explains a and a explains no object.

We can also use model finders to perform 'sanity checks': We instruct Nitpick to find a countermodel for some specifically tailored formula which we want to make sure is not valid, because of its implausibility from the point of view of the author (as we interpret him). We check below, for instance, that our axioms are not too strong as to imply *metaphysical necessitism* (i.e. that all beings necessarily exist) or *modal collapse* (i.e. that all truths are necessary). Since both would trivially validate the argument.

lemma $\lfloor \forall x.\ E!\ x \rfloor$
  nitpick[*user-axioms*] **oops** — Countermodel found: necessitism is not valid
lemma $\lfloor \varphi \to \Box\varphi \rfloor$
  nitpick[*user-axioms*] **oops** — Countermodel found: modal collapse is not valid

Model finders like Nitpick are able to verify consistency (by finding a model) or non-validity (by finding a countermodel) for a given formula. When it comes to verifying validity or invalidity, we are use automated theorem provers. Isabelle comes with various different provers tailored for specific kinds of problems and thus employing different approaches, strategies and heuristics. We typically make extensive use of Isabelle's *Sledgehammer* tool [10], which integrates several state-of-the-art external theorem provers and feeds them with different combinations of axioms and the conjecture in question. If successful, *Sledgehammer* returns valuable dependency information (the exactly required axioms and definitions to prove a given conjecture) back to Isabelle, which then exploits this information to (re-)construct a trusted proof with own, internal proof automation means. The entire process often only takes a few seconds.

By using Sledgehammer we can here verify the validity of our partial conclusions (C1, C5 and C7) and even find the premises they rely upon.[27]

---

[27] We prove theorems in Isabelle here by using the keyword "by" followed by the name of an Isabelle-internal and thus trusted proof method (generally, some computer-implemented algorithm). Some methods commonly used in Isabelle are: *simp* (term rewriting), *blast* (tableaus), *meson* (model elimination), *metis* (ordered resolution and paramodulation) and *auto* (classical reasoning and term rewriting). As explained, these methods were automatically suggested and applied by the Sledgehammer tool. The interactive user in fact does not need to know, or learn, much about these methods in the beginning (he will benefit a lot though, if he does).

(C1) *All abstract beings depend for their existence on concrete beings.*

theorem $C1$: $\lfloor \forall^A x.\ Abstract\ x \to (\exists y.\ Concrete\ y \land x\ dependsOn\ y) \rfloor$
  using $P3\ P4$ **by** *blast*

(C5) *In every possible world there exist concrete beings.*

theorem $C5$: $\lfloor \exists^A x.\ Concrete\ x \rfloor$
  using $P2\ P3\ P4$ **by** *blast*

(C7) *The existence of necessary abstract beings needs to be explained.*

theorem $C7$: $\lfloor \forall^A x.\ (Necessary\ x \land Abstract\ x) \to Explained\ x \rfloor$
  using $P3\ P6$ **by** *simp*

The last three conclusions are shown by Nitpick to be non-valid even in the stronger *S5* logic. *S5* can be easily introduced by postulating that the reachability relation R is an equivalence relation. This exploits the *Sahlqvist correspondence* which relates modal axioms to constraints on a model's reachability relation: reflexivity, symmetry, seriality, transitivity and euclideanness imply axioms T, B, D, IV, V respectively (and also the other way round).

axiomatization where
  $S5$: *equivalence* R — We assume T: $\Box\varphi \to \varphi$, B: $\varphi \to \Box\Diamond\varphi$ and 4: $\Box\varphi \to \Box\Box\varphi$

(C8) *The existence of necessary abstract beings can only be explained by concrete beings.*

lemma $C8$: $\lfloor \forall^A x.(Necessary\ x \land Abstract\ x) \to (\forall^A y.\ y\ explains\ x \to Concrete\ y) \rfloor$
  nitpick[*user-axioms*] **oops**

(C9) *The existence of necessary abstract beings is explained by one or more necessary concrete (Godlike) beings.*

lemma $C9$: $\lfloor \forall^A x.(Necessary\ x \land Abstract\ x) \to (\exists^A y.\ y\ explains\ x \land Godlike\ y) \rfloor$
  nitpick[*user-axioms*] **oops**

(C10) *A necessary concrete (Godlike) being exists.*

theorem $C10$: $\lfloor \exists^A x.\ Godlike\ x \rfloor$ **nitpick**[*user-axioms*] **oops**

Note that Nitpick does not only spare us the effort of searching for non-existent proofs but also provides us with very helpful information when it comes to fix an argument by giving us a text-based description of the (counter-)model found. We present below another screenshot showing Nitpick's counterexample for C10:

```
267 (** The existence of necessary abstract beings can only be expl
268 theorem C8: "⌊∀ᴬx. (Necessary x ∧ Abstract x) → (∀ᴬy. y explains x → (
269 (** The existence of necessary abstract beings is explained by one or m
270 theorem C9: "⌊∀ᴬx. (Necessary x ∧ Abstract x) → (∃ᴬy. y explains x ∧ G
271 (** A necessary concrete [Godlike] being exists *)
272 theorem C10:    "⌊∃ᴬx. Godlike x⌋" nitpick[user_axioms] oops
```

Nitpick found a counterexample for card e = 3 and card w = 2:

  Skolem constants:
    λx. ??.Necessary.v = (λx. _)(a := w₁, b := w₁, c := w₂)
    w = w₂
  Constants:
    op actualizedAt =
      (λx. _)
      ((a, w₁) := False, (a, w₂) := True, (b, w₁) := True,
       (b, w₂) := True, (c, w₁) := True, (c, w₂) := True)
    op r =
      (λx. _)
      ((w₁, w₁) := False, (w₁, w₂) := False, (w₂, w₁) := True,
       (w₂, w₂) := False)
    Concrete =
      (λx. _)
      ((a, w₁) := False, (a, w₂) := True, (b, w₁) := False,

By employing the Isabelle proof assistant we have proven non-valid a first formalization attempt of Lowe's modal ontological argument. This is, however, just the first of many series of iterations in our interpretive endeavor. Based on the information recollected so far, we can proceed to make the adjustments necessary to validate the argument. We will see how these adjustments have an impact on the inferential role of all concepts (necessariness, concreteness, dependence, explanation, etc.) and therefore on their meaning.

Second Iteration Series: Validating the Argument I

By carefully examining the above countermodel for C10, it has been noticed that some necessary beings, which are abstract in the actual world, may indeed be concrete in other reachable worlds. Lowe has previously presented numbers as an example of such necessary abstract beings. It can be argued that numbers, while existing necessarily, can never be concrete in any possible world, so we add the restriction of abstractness being an essential property, i.e. a locally rigid predicate.

axiomatization where
  abstractness-essential: ⌊∀ x. Abstract x → □Abstract x⌋

theorem $C10$: $\lfloor \exists^A x.\ Godlike\ x \rfloor$
  nitpick[*user-axioms*] **oops** — Countermodel found

Again, we have used model finder Nitpick to get a counterexample for C10, so the former restriction is not enough to prove this conclusion. We try postulating further restrictions on the reachability relation R, which, taken together, would amount to it being an equivalence relation. This would make for a modal logic $S5$ (see *Sahlqvist correspondence*), and thus the abstractness property becomes a (globally) rigid predicate.

axiomatization where
  T-axiom: *reflexive* R **and** — $\Box\varphi \to \varphi$
  B-axiom: *symmetric* R **and** — $\varphi \to \Box\Diamond\varphi$
  IV-axiom: *transitive* R — $\Box\varphi \to \Box\Box\varphi$

theorem $C10$: $\lfloor \exists^A x.\ Godlike\ x \rfloor$
  nitpick[*user-axioms*] **oops** — Countermodel found

By examining the new countermodel found by Nitpick, we noticed that at some worlds there are non-existent concrete beings. We want to disallow this possibility, so we make concreteness an existence-entailing property.

axiomatization where *concrete-exist*: $\lfloor \forall x.\ Concrete\ x \to E!\ x \rfloor$

We carry out the usual 'sanity checks' to make sure the argument has not become trivialized.[28]

lemma *True*
  nitpick[*satisfy*, *user-axioms*] **oops** — Model found: axioms are consistent
lemma $\lfloor \forall x.\ E!\ x \rfloor$
  nitpick[*user-axioms*] **oops** — Countermodel found: necessitism is not valid
lemma $\lfloor \varphi \to \Box\varphi \rfloor$
  nitpick[*user-axioms*] **oops** — Countermodel found: modal collapse is not valid

Since Nitpick could not find a countermodel for C10, we have enough confidence in its validity to ask another automated reasoning tool: Isabelle's *Sledgehammer* [10] to search for a proof.

theorem $C10$: $\lfloor \exists^A x.\ Godlike\ x \rfloor$ **using** *Existence-def Necessary-def*
  abstractness-essential concrete-exist P2 C1 B-axiom **by** *meson*

Sledgehammer is able to find a proof relying on all premises but the two modal axioms T and IV. Thus, by the end of this series of iterations, we have seen that

---

[28] These checks are constantly carried out after postulating axioms for every iteration, so we won't mention them anymore.

Lowe's modal ontological argument depends for its validity on three unstated (i.e. implicit) premises: the essentiality of abstractness, the existence-entailing nature of concreteness, and the modal axiom B ($\varphi \to \Box\Diamond\varphi$). Moreover, we shed some light on the meaning of the concepts of abstractness and concreteness, as we disclose further premises which shape their inferential role in the argument.

Third Iteration Series: Validating the Argument II

In this iteration series we want to explore the critical potential of computational hermeneutics. In this slightly simplified variant (without the implicit premises stated in the previous version), premises P1 to P5 remain unchanged, while none of the last three premises (P6 to P8) show up anymore. Those last premises have been introduced by Lowe in order to interrelate the concepts of explanation and dependence in such a way that they play somewhat opposite roles, without one being the inverse of the other. Nonetheless, we will go all the way and assume that explanation and dependence are indeed inverse relations, for we want to understand how the interrelation of these two concepts affects the validity of the argument.

axiomatization where
  dep-expl-inverse: $\lfloor \forall\, x\, y.\ y\ explains\ x \leftrightarrow x\ dependsOn\ y \rfloor$

Let us first prove the relevant partial conclusions.

theorem *C1*: $\lfloor \forall^A x.\ Abstract\ x \to (\exists\, y.\ Concrete\ y \wedge x\ dependsOn\ y) \rfloor$
  using *P3 P4* **by** *blast*

theorem *C5*: $\lfloor \exists^A x.\ Concrete\ x \rfloor$
  using *P2 P3 P4* **by** *blast*

theorem *C7*: $\lfloor \forall^A x.\ (Necessary\ x \wedge Abstract\ x) \to Explained\ x \rfloor$
  using *Explained-def P3 P4 dep-expl-inverse* **by** *meson*

However, the conclusion C10 is still countersatisfiable, as shown by Nitpick.

theorem *C10*: $\lfloor \exists^A x.\ Godlike\ x \rfloor$
  nitpick[*user-axioms*] **oops** — Countermodel found

Next, let us try assuming a stronger modal logic. We can do this by postulating further modal axioms using the *Sahlqvist correspondence* and asking Sledgehammer to find a proof. Sledgehammer is in fact able to find a proof for C10 which only relies on the modal axiom T ($\Box\varphi \to \varphi$).

axiomatization where
  T-axiom: *reflexive R* **and** — $\Box\varphi \to \varphi$

B-axiom: *symmetric R* **and** $\;-\varphi \to \Box\Diamond\varphi$
IV-axiom: *transitive R* $\quad -\Box\varphi \to \Box\Box\varphi$

theorem *C10*: $\lfloor \exists^A x.\ Godlike\ x \rfloor$ **using** *Contingent-def Existence-def*
  P2 P3 P4 P5 dep-expl-inverse T-axiom **by** *meson*

In this series of iterations we have verified a modified version of the original argument by Lowe. Our understanding of the concepts of *ontological dependence* and *metaphysical explanation* (in the context of Lowe's argument) has changed after the introduction of an additional axiom constraining both: they are now inverse relations. This new understanding of the inferential role of the above concepts of dependence and explanation has been reached on the condition that the ontological argument, as stated in natural language, must hold (in accordance to the *principle of charity*). Depending on our stance on this matter, we may either feel satisfied with this result or want to consider further alternatives. In the former case we would have reached a state of *reflective equilibrium*. In the latter we would rather carry on with our iterative process in order to further illuminate the meaning of the expressions involved in this argument.

Fourth Iteration Series: Simplifying the Argument

After some further iterations we arrive at a new variant of Lowe's argument: Premises P1 to P4 remain unchanged and a new premise D5 ("x depends for its existence on y := necessarily, x exists only if y exists") is added. D5 corresponds to the 'definition' of ontological dependence as put forth by Lowe in his article (though only for illustrative purposes). As mentioned before, this purported definition was never meant by him to become part of the argument. Nevertheless, we show here how, by assuming the left-to-right direction of this definition, we get in a position to prove the main conclusions without any further assumptions.

axiomatization where *D5*: $\lfloor \forall^A x\ y.\ x\ dependsOn\ y \to \Box(E!\ x \to E!\ y) \rfloor$

theorem *C1*: $\lfloor \forall^A x.\ Abstract\ x \to (\exists y.\ Concrete\ y \land x\ dependsOn\ y) \rfloor$
  using *P3 P4* **by** *meson*

theorem *C5*: $\lfloor \exists^A x.\ Concrete\ x \rfloor$ **using** *P2 P3 P4* **by** *meson*

theorem *C10*: $\lfloor \exists^A x.\ Godlike\ x \rfloor$
  using *Necessary-def P2 P3 P4 D5* **by** *meson*

In this variant, we have been able to verify the conclusion of the argument without appealing to the concept of metaphysical explanation. We were able to get by with just the concept of ontological dependence by explicating it in terms of existence and necessity (as suggested by Lowe).

As a side note, we can also prove that the original premise P5 ("No contingent being can explain the existence of a necessary being") directly follows from D5 by redefining metaphysical explanation as the inverse relation of ontological dependence.

abbreviation *explanation*::($e \Rightarrow e \Rightarrow wo$) (**infix** *explains*)
  where *y explains x* $\equiv$ *x dependsOn y*

lemma *P5*: $\lfloor \neg (\exists^A x.\ \exists^A y.\ Contingent\ y \wedge Necessary\ x \wedge y\ explains\ x) \rfloor$
  using *Necessary-def Contingent-def D5* **by** *meson*

In this series of iterations we have reworked Lowe's argument so as to get rid of the somewhat obscure concept of metaphysical explanation, thus simplifying the argument. We also got some insight into Lowe's concept of ontological dependence vis-à-vis its inferential role in the argument (by axiomatizing its relation with the concepts of existence and necessity in D5).

There are still some interesting issues to consider. Note that the definitions of existence and being-dependent (axioms "Existence-def" and "Dependent-def" respectively) are not needed in any of the highly optimized proofs found by our automated tools. This raises some suspicions concerning the role played by the existence predicate in the definitions of necessariness and contingency, as well as putting into question the need for a definition of being-dependent linked to the ontological dependence relation. We will see in the following section that our suspicions are justified and that this argument can be dramatically simplified.

Fifth Iteration Series: Arriving at a Non-Modal Argument

In the next iterations, we want to explore once again the critical potential of computational hermeneutics by challenging another of the author's claims: that this argument is a *modal* one. A new simplified version of Lowe's argument is obtained after abandoning the concept of existence altogether and redefining necessariness and contingency accordingly. As we will see, this variant is actually non-modal and can be easily formalized in first-order predicate logic.

A more literal reading of Lowe's article has suggested a simplified formalization, in which necessariness and contingency are taken as complementary predicates. According to this, our domain of discourse becomes divided in four main categories, as exemplified in the table below.[29]

---

[29] As Lowe explains in the article, "there is no logical restriction on combinations of the properties

	Abstract	Concrete
Necessary	Numbers	God
Contingent	Fiction	Stuff

consts *Necessary*::$e{\Rightarrow}wo$
abbreviation *Contingent*::$e{\Rightarrow}wo$ **where** *Contingent* $x \equiv \neg(\textit{Necessary } x)$

consts *Concrete*::$e{\Rightarrow}wo$
abbreviation *Abstract*::$e{\Rightarrow}wo$ **where** *Abstract* $x \equiv \neg(\textit{Concrete } x)$

abbreviation *Godlike*::$e{\Rightarrow}w{\Rightarrow}bool$ **where** *Godlike* $x \equiv \textit{Necessary } x \wedge \textit{Concrete } x$

consts *dependence*::$e{\Rightarrow}e{\Rightarrow}wo$ (**infix** *dependsOn*)
abbreviation *explanation*::$(e{\Rightarrow}e{\Rightarrow}wo)$ (**infix** *explains*)
  where y *explains* $x \equiv x$ *dependsOn* y

As shown below, we can even define being-dependent as a *primitive* predicate (i.e. bearing no relation to ontological dependence) and still be able to validate the argument. Being-independent is defined as the negation of being-dependent.

consts *Dependent*::$e{\Rightarrow}wo$
abbreviation *Independent*::$e{\Rightarrow}wo$ **where** *Independent* $x \equiv \neg(\textit{Dependent } x)$

By taking, once again, metaphysical explanation as the inverse relation of ontological dependence and by assuming premises P2 to P5 we can prove conclusion C10.

axiomatization where
P2: $\lfloor \exists x.\ \textit{Necessary } x \wedge \textit{Abstract } x \rfloor$ **and**
P3: $\lfloor \forall x.\ \textit{Abstract } x \rightarrow \textit{Dependent } x \rfloor$ **and**
P4: $\lfloor \forall x.\ \textit{Dependent } x \rightarrow (\exists y.\ \textit{Independent } y \wedge x\ \textit{dependsOn } y) \rfloor$ **and**
P5: $\lfloor \neg(\exists x.\ \exists y.\ \textit{Contingent } y \wedge \textit{Necessary } x \wedge y\ \textit{explains } x) \rfloor$

theorem *C10*: $\lfloor \exists x.\ \textit{Godlike } x \rfloor$ **using** *P2 P3 P4 P5* **by** *blast*

Note that, in the axioms above, all restricted (actualist) quantifiers have been changed into unrestricted (possibilist) quantifiers, following the elimination of the concept of existence from our argument: Our quantifiers now range over all beings, because all beings exist. Also note that modal operators have disappeared; thus, this new variant is directly formalizable in classical first-order logic.

---

involved in the concrete/abstract and the necessary/contingent distinctions. In principle, then, we can have contingent concrete beings, contingent abstract beings, necessary concrete beings, and necessary abstract beings."

Sixth Iteration Series: Modified Modal Argument I

In the following two series of iterations, we want to illustrate the use of the *computational hermeneutics* approach in those cases where we must start our interpretive endeavor with no *explicit* understanding of the concepts involved. In such cases, we start by taking all concepts as primitive without stating any definition explicitly. We will see how we gradually improve our understanding of these concepts in the iterative process of adding and removing axioms, thus framing their inferential role in the argument.

consts *Concrete*::$e \Rightarrow wo$
consts *Abstract*::$e \Rightarrow wo$
consts *Necessary*::$e \Rightarrow wo$
consts *Contingent*::$e \Rightarrow wo$
consts *dependence*::$e \Rightarrow e \Rightarrow wo$ (**infix** *dependsOn*)
consts *explanation*::$e \Rightarrow e \Rightarrow wo$ (**infix** *explains*)
consts *Dependent*::$e \Rightarrow wo$
abbreviation *Independent*::$e \Rightarrow wo$ **where** *Independent* $x \equiv \neg(Dependent\ x)$

In order to honor the original intention of the author, i.e., providing a *modal* variant of St. Anselm's ontological argument, we are required to make a change in Lowe's original formulation. In this variant we will restate the expressions "necessary abstract" and "necessary concrete" as "necessari*ly* abstract" and "necessari*ly* concrete" respectively. With this new adverbial reading we are no longer talking about the concept of *necessariness*, but of *necessity* instead, so we use the modal box operator ($\Box$) for its formalization. It can be argued that in this variant we are not concerned with the interpretation of the *original* natural-language argument anymore. We are rather interested in showing how the computational hermeneutics method can go beyond simple interpretation and foster a creative approach to assessing and improving philosophical arguments.

Premise P1 now reads: "God is, by definition, a necessari*ly* concrete being."

abbreviation *Godlike*::$e \Rightarrow wo$ **where** *Godlike* $x \equiv \Box Concrete\ x$

Premise P2 reads: "Some necessari*ly* abstract beings exist". The rest of the premises remains unchanged.

axiomatization where
$P2$: $\lfloor \exists x.\ \Box Abstract\ x \rfloor$ **and**
$P3$: $\lfloor \forall x.\ Abstract\ x \rightarrow Dependent\ x \rfloor$ **and**
$P4$: $\lfloor \forall x.\ Dependent\ x \rightarrow (\exists y.\ Independent\ y \wedge x\ dependsOn\ y) \rfloor$ **and**
$P5$: $\lfloor \neg(\exists x.\ \exists y.\ Contingent\ y \wedge Necessary\ x \wedge y\ explains\ x) \rfloor$

Without postulating any additional axioms, C10 ("A *necessarily* concrete being exists") can be falsified by Nitpick.

theorem *C10*: ⌊∃ x. Godlike x⌋
  nitpick oops — Countermodel found

An explication of the concepts of necessariness, contingency and explanation is provided below by axiomatizing their interrelation to other concepts. We will now regard necessariness as being *necessarily abstract* or *necessarily concrete*, and explanation as the inverse relation of dependence, as before.

axiomatization where
  Necessary-expl: ⌊∀ x. Necessary x ↔ (□Abstract x ∨ □Concrete x)⌋ **and**
  Contingent-expl: ⌊∀ x. Contingent x ↔ ¬Necessary x⌋ **and**
  Explanation-expl: ⌊∀ x y. y explains x ↔ x dependsOn y⌋

Without any further constraints, C10 becomes again falsified by Nitpick.

theorem *C10*: ⌊∃ x. Godlike x⌋
  nitpick oops — Countermodel found

We postulate further modal axioms (using the *Sahlqvist correspondence*) and ask Isabelle's Sledgehammer tool for a proof. Sledgehammer is able to find a proof for C10 which only relies on the modal axiom T ($\Box\varphi \to \varphi$).

axiomatization where
  T-axiom: *reflexive R* **and** — $\Box\varphi \to \varphi$
  B-axiom: *symmetric R* **and** — $\varphi \to \Box\Diamond\varphi$
  IV-axiom: *transitive R* — $\Box\varphi \to \Box\Box\varphi$

theorem *C10*: ⌊∃ x. Godlike x⌋ **using** *Contingent-expl Explanation-expl*
  Necessary-expl P2 P3 P4 P5 T-axiom **by** *metis*

Seventh Iteration Series: Modified Modal Argument II

As in the previous variant, we will illustrate here how the meaning (as inferential role) of the expressions involved in the argument gradually becomes explicit in the process of axiomatizing further constraints. We follow on with the adverbial reading of the expression "necessary" but provide an improved explication of necessariness (and contingency). We think that this explication, in comparison to the previous one, better fits our intuitive pre-understanding of the concept of being a necessary (or contingent) being. Thus, we will now regard necessariness as being *necessarily* abstract or concrete. (As before, we regard here metaphysical explanation as the inverse of the ontological dependence relation.)

axiomatization where
Necessary-expl: $\lfloor \forall x.\ Necessary\ x \leftrightarrow \square(Abstract\ x \vee Concrete\ x) \rfloor$ and
Contingent-expl: $\lfloor \forall x.\ Contingent\ x \leftrightarrow \neg Necessary\ x \rfloor$ and
Explanation-expl: $\lfloor \forall x\ y.\ y\ explains\ x \leftrightarrow x\ dependsOn\ y \rfloor$

These constraints are, however, not enough to ensure the argument's validity, as confirmed by Nitpick.

theorem *C10*: $\lfloor \exists x.\ Godlike\ x \rfloor$ **nitpick oops** — Countermodel found

After some iterations, we see that, by giving a more satisfactory explication of the concept of necesariness, we are also required to (i) assume the essentiality of abstractness (as we did in a former iteration), and (ii) restrict the reachability relation by enforcing its symmetry (i.e. assuming the modal axiom B).

axiomatization where
abstractness-essential: $\lfloor \forall x.\ Abstract\ x \rightarrow \square Abstract\ x \rfloor$ and
B-Axiom: symmetric R — $\varphi \rightarrow \square \diamond \varphi$

theorem *C10*: $\lfloor \exists x.\ Godlike\ x \rfloor$ **using** *Contingent-expl Explanation-expl Necessary-expl P2 P3 P4 P5 abstractness-essential B-Axiom* **by** *metis*

In each of the previous versions we have seen how our understanding of the concepts of being-necessary (necessariness), being-contingent (contingency), explanation, dependence, abstractness, concreteness, etc. has gradually evolved thanks to the iterative holistic method made possible by the real-time feedback provided by Isabelle's automated proving tools.

We think that, after this last series of iterations, the use of the *computational hermeneutics* method has been illustrated adequately. We do not claim that this formalization of Lowe's argument is its best or most adequate one; it is just a consequence of the path we have followed by coming up with new ideas and testing them with the help of automated tools. In our view, while the third variant may be the closest one to Lowe's original formulation, it is this latter (seventh) variant the one which strikes the best balance between interpretation and critical assessment of this argument. We encourage the reader to continue with this process until arriving to his/her own *reflective equilibrium* (possibly by building upon our computer-verified work [22] available at the *Archive of Formal Proofs*).[30]

---

[30]The Archive of Formal Proofs (www.isa-afp.org) is a collection of proof libraries, examples, and larger scientific developments, mechanically checked using the Isabelle proof assistant. It is organized in the way of a scientific journal and submissions are refereed.

Conclusion

We have argued for the role of formal logic as an *ars explicandi* and the possibility of applying it to foster our understanding of rational arguments (in particular metaphysical and theological ones). We understand the give-and-take process aiming at an adequate formal reconstruction of a natural-language argument in itself as a kind of interpretive endeavor. Moreover, we have argued that, by using automated reasoning technology to systematically explore the many different inferential possibilities latent in a formalized argument, we can make explicit the inferential role played by its constituent expressions and thus better understand their meaning in the given interpretation context.

As a computer-assisted method, computational hermeneutics aims at complementing our human ingenuity with the data-processing power of modern computers and at using this synergy to make interpretation more effective. In a similar vein, we currently work on how to apply this approach in the computer science field of *natural language understanding*. Specifically, we want to tackle the problem of formalization: how to search *methodically* for the most appropriate logical form(s) of a given natural-language argument, by casting its individual statements into expressions of some sufficiently expressive logical language. Being able to automatically extract a formal representation for some piece of natural-language discourse, by taking into account its holistically-determined logical location in a web of possible inferences, is an important step towards the deep semantic analysis and critical assessment of non-trivial natural-language discourse. Further applications in areas like knowledge/ontology extraction, semantic web and legal informatics are currently being contemplated.

References

[1] J. Alama, P. E. Oppenheimer, and E. N. Zalta. Automating Leibniz's theory of concepts. In A. P. Felty and A. Middeldorp, editors, *Automated Deduction - CADE-25 - 25th International Conference on Automated Deduction, Berlin, Germany, August 1-7, 2015, Proceedings*, volume 9195 of *LNCS*, pages 73–97. Springer, 2015.

[2] C. Baumberger and G. Brun. Dimensions of objectual understanding. *Explaining understanding. New perspectives from epistemology and philosophy of science*, pages 165–189, 2016.

[3] M. Baumgartner and T. Lampert. Adequate formalization. *Synthese*, 164(1):93–115, 2008.

[4] C. Benzmüller. A top-down approach to combining logics. In J. Filipe and A. Fred, editors, *Proc. of the 5th International Conference on Agents and Artificial Intelligence (ICAART)*, volume 1, pages 346–351, Barcelona, Spain, 2013. SCITEPRESS – Science and Technology Publications, Lda.

[5] C. Benzmüller. Recent successes with a meta-logical approach to universal logical reasoning (extended abstract). In S. A. da Costa Cavalheiro and J. L. Fiadeiro, editors, *Formal Methods: Foundations and Applications - 20th Brazilian Symposium, SBMF 2017, Recife, Brazil, November 29 - December 1, 2017, Proceedings*, volume 10623 of *Lecture Notes in Computer Science*, pages 7–11. Springer, 2017.

[6] C. Benzmüller and L. Paulson. Quantified multimodal logics in simple type theory. *Logica Universalis (Special Issue on Multimodal Logics)*, 7(1):7–20, 2013.

[7] C. Benzmüller, L. Weber, and B. Woltzenlogel Paleo. Computer-assisted analysis of the Anderson-Hájek controversy. *Logica Universalis*, 11(1):139–151, 2017.

[8] C. Benzmüller and B. Woltzenlogel Paleo. Automating Gödel's ontological proof of God's existence with higher-order automated theorem provers. In T. Schaub, G. Friedrich, and B. O'Sullivan, editors, *ECAI 2014*, volume 263 of *Frontiers in Artificial Intelligence and Applications*, pages 93 – 98. IOS Press, 2014.

[9] C. Benzmüller and B. Woltzenlogel Paleo. The inconsistency in Gödel's ontological argument: A success story for AI in metaphysics. In *IJCAI 2016*, 2016.

[10] J. Blanchette, S. Böhme, and L. Paulson. Extending Sledgehammer with SMT solvers. *Journal of Automated Reasoning*, 51(1):109–128, 2013.

[11] J. Blanchette and T. Nipkow. Nitpick: A counterexample generator for higher-order logic based on a relational model finder. In *Proc. of ITP 2010*, volume 6172 of *LNCS*, pages 131–146. Springer, 2010.

[12] N. Block. Semantics, conceptual role. In *Routledge Encyclopedia of Philosophy*. Taylor and Francis, 1998.

[13] R. B. Brandom. *Making It Explicit: Reasoning, Representing, and Discursive Commitment*. Harvard University Press, 1994.

[14] G. Brun. Reconstructing arguments. formalization and reflective equilibrium. *Logical analysis and history of philosophy*, 17:94–129, 2014.

[15] D. Davidson. Radical interpretation interpreted. *Philosophical Perspectives*, 8:121–128, January 1994.

[16] D. Davidson. *Essays on actions and events: Philosophical essays*, volume 1. Oxford University Press on Demand, 2001.

[17] D. Davidson. On the very idea of a conceptual scheme. In *Inquiries into Truth and Interpretation*. Oxford University Press, September 2001.

[18] D. Davidson. Radical interpretation. In *Inquiries into Truth and Interpretation*. Oxford University Press, September 2001.

[19] G. Eder and E. Ramharter. Formal reconstructions of St. Anselm's ontological argument. *Synthese: An International Journal for Epistemology, Methodology and Philosophy of Science*, 192(9), October 2015.

[20] C. Elgin. *Considered judgment*. Princeton University Press, 1999.

[21] D. Fuenmayor and C. Benzmüller. Automating emendations of the ontological argument in intensional higher-order modal logic. In *KI 2017: Advances in Artificial Intelligence 40th Annual German Conference on AI, Dortmund, Germany, September 25-29, 2017, Proceedings*, volume 10505 of *LNAI*, pages 114–127. Springer, 2017.

[22] D. Fuenmayor and C. Benzmüller. Computer-assisted reconstruction and assessment of E. J. Lowe's modal ontological argument. *Archive of Formal Proofs*, Sept. 2017. http://isa-afp.org/entries/Lowe_Ontological_Argument.html, Formal proof development.

[23] D. Fuenmayor and C. Benzmüller. Types, Tableaus and Gödel's God in Isabelle/HOL. *Archive of Formal Proofs*, May 2017. http://isa-afp.org/entries/Types_Tableaus_and_Goedels_God.html, Formal proof development.

[24] T. F. Godlove, Jr. *Religion, Interpretation and Diversity of Belief: The Framework Model from Kant to Durkheim to Davidson*. Cambridge University Press, 1989.

[25] T. F. Godlove, Jr. Saving belief: on the new materialism in religious studies. In N. Frankenberry, editor, *Radical interpretation in religion*. Cambridge University Press, 2002.

[26] T. Hales, M. Adams, G. Bauer, T. D. Dang, J. Harrison, L. T. Hoang, C. Kaliszyk, V. Magron, S. Mclaughlin, T. Nguyen, and et al. A formal proof of the kepler conjecture. *Forum of Mathematics, Pi*, 5, 2017.

[27] G. Harman. (Nonsolipsistic) conceptual role semantics. In E. Lepore, editor, *Notre Dame Journal of Formal Logic*, pages 242–256. Academic Press, 1987.

[28] P. Horwich. *Meaning*. Oxford University Press, 1998.

[29] E. J. Lowe. Ontological dependence. In E. N. Zalta, editor, *The Stanford Encyclopedia of Philosophy*. Metaphysics Research Lab, Stanford University, spring 2010 edition, 2010.

[30] E. J. Lowe. A modal version of the ontological argument. In J. P. Moreland, K. A. Sweis, and C. V. Meister, editors, *Debating Christian Theism*, chapter 4, pages 61–71. Oxford University Press, 2013.

[31] T. Nipkow, L. C. Paulson, and M. Wenzel. *Isabelle/HOL — A Proof Assistant for Higher-Order Logic*. Number 2283 in LNCS. Springer, 2002.

[32] P. Oppenheimer and E. Zalta. A computationally-discovered simplification of the ontological argument. *Australasian Journal of Philosophy*, 89(2):333–349, 2011.

[33] P. Pagin. Is compositionality compatible with holism? *Mind & Language*, 12(1):11–33, March 1997.

[34] P. Pagin. Meaning holism. In E. Lepore, editor, *The Oxford handbook of philosophy of language*. Oxford University Press, 1. publ. in paperback edition, 2008.

[35] F. J. Pelletier. Holism and compositionality. In W. Hinzen, E. Machery, and M. Werning, editors, *The Oxford Handbook of Compositionality*. Oxford University Press, 1 edition, February 2012.

[36] F. J. Pelletier, G. Sutcliffe, and C. Suttner. The development of CASC. *AI Commun.*,

15(2,3):79–90, Aug. 2002.

[37] J. Peregrin. *Inferentialism: Why rules matter*. Springer, 2014.

[38] J. Peregrin and V. Svoboda. Criteria for logical formalization. *Synthese*, 190(14):2897–2924, 2013.

[39] J. Peregrin and V. Svoboda. *Reflective Equilibrium and the Principles of Logical Analysis: Understanding the Laws of Logic*. Routledge Studies in Contemporary Philosophy. Taylor and Francis, 2017.

[40] F. Portoraro. Automated reasoning. In E. N. Zalta, editor, *The Stanford Encyclopedia of Philosophy*. Metaphysics Research Lab, Stanford University, winter 2014 edition, 2014.

[41] J. Rushby. The ontological argument in PVS. In *Proc. of CAV Workshop "Fun With Formal Methods"*, St. Petersburg, Russia, 2013.

[42] G. Sutcliffe and C. Suttner. The TPTP problem library. *Journal of Automated Reasoning*, 21(2):177–203, Oct 1998.

[43] A. Tarski. The concept of truth in formalized languages. *Logic, semantics, metamathematics*, 2:152–278, 1956.

[44] F. Wiedijk. *The Seventeen Provers of the World: Foreword by Dana S. Scott (Lecture Notes in Computer Science / Lecture Notes in Artificial Intelligence)*. Springer-Verlag New York, Inc., Secaucus, NJ, USA, 2006.

[45] M. Williams. Meaning and deflationary truth. *Journal of philosophy*, XCVI(11):545–564, November 1999.

[46] T. Williamson. *Modal Logic as Metaphysics*. Oxford University Press, 2013.

www.ingramcontent.com/pod-product-compliance
Lightning Source LLC
Chambersburg PA
CBHW080439110426
42743CB00016B/3211